ME CARIBOU IS ON FIRE!

INTERNATIONAL ADVENTURES
OF AN ALASKAN HUNTING GUIDE

PETE BUIST

MASTER GUIDE #79 (RETIRED)

PO Box 221974 Anchorage, Alaska 99522-1974
books@publicationconsultants.com—www.publicationconsultants.com

ISBN Number: 978-1-63747-055-8
ISBN eBook: 978-1-63747-056-5

Library of Congress Catalog Card Number: 2021918588

Copyright 2021 Pete Buist
—First Edition 2021—

All rights reserved, including the right of reproduction in any form, or by any mechanical or electronic means including photocopying or recording, or by any information storage or retrieval system, in whole or in part in any form, and in any case not without the written permission of the author and publisher.

"Uncertain Shores"

The painting featured on the cover, "Uncertain Shores," is by Marc Lee, a forester friend from Fairbanks. It depicts migrating caribou crossing the Kobuk River in northwest Alaska. Limited edition prints of the 16" x 32" painting are available from the artist. **Contact:** denaliprints@gmail.com.

Manufactured in the United States of America

Dedication

This book is dedicated to all those outdoor minded kids (both boys and girls) in small, flat states who read books (and nowadays, surf the Internet) and dream of heading off to the forests, swamps, savannahs, kopjes, mountains and glaciers of the world's great game fields to pursue wonderful creatures.

There is no reason you cannot do it. In fact you can do it! I've seen ME do it!

Pete Buist
Fairbanks, Alaska

Acknowledgments

I had the encouragement of lots of folks. My wife Emma Lee and both boys, Jason and Peyton kept at me to do a book. Lots of people who were avid readers of my journals over the years kept saying "you should write a book." (More people said that than said "I'd buy a copy!" but I digress...) A myriad of friends in hunting camps all over the world kept the pressure on me to finally "just write it and be done with it!" The most exuberant reader of my journals was Bill Miller Sr. of Llano, Texas. Bill died in 2009 soon after he and I got back from a hunt in New Zealand. Bill never let up on me about converting those journals to a book. He was on my mind 10 years later when I helped his 12 year old granddaughter Shelby take her first black bear. Much of the success I have had at living a life that consists of a series of adventures are due to the encouragement and teachings of a lot of people. The most significant of these are described in more detail in the chapter on "mentors." There are a ton of others.

I would be totally remiss if I did not give a collective nod to the many clients who trusted me to help them achieve their Alaska hunting dreams. I achieved many of my dreams and I felt it was an honorable exercise to help others do the same. I can honestly say that I learned something from every single client. Hopefully they learned things from me as well. They all were darned good at what they did and successful enough to be able to come to Alaska. We have had some amazing conversations around wood stoves and campfires over the years!

A special mention needs to go several old-time guides. What an honor it has been to share campfires and coffee (as well as legislative conference rooms) with them. The Fairbanks guys of note include Bud Helmericks, Sam Snyder, Urban Rahoi, Dan Wetzel, Lynn Castle, Joe Want, Ken Fanning, Bernd Gaedake, Peter Merry, Chuck Gray and Pete Shepherd. In addition to sharing Alaska lore with me, Bill Waugaman Sr. talked me into getting off my duff and going to Africa! Specifically, Bill encouraged me to hunt kudu. A fine kudu was the very first African animal that I took on my very first safari.

Ironically, Bill passed away while I was in Africa on that trip. If only I could have told him directly how right he was! Guides from elsewhere in Alaska but who were influential include Clark Engle, Tom Walker, Stan Frost, Keith Johnson, Ray McNutt, Leon Francisco, George Palmer, Phil Driver, Dick Gunlogson, Mel Gillis, Dick Rohrer and Bud Branham. I've probably forgotten some and for that, I apologize.

Of course I learned useful things from other guides who shall remain nameless... during disciplinary hearings before the old Guide Licensing & Control Board and later the Alaska Big Game Commercial Services Board (on which I still sit.) It's OK to learn what NOT to do too!

In the course of living in Fairbanks and hunting around Alaska, I came to know many fine Alaska Wildlife Troopers. When I started they were known as Fish and Wildlife Protection officers. I have a ton of respect for a number of them. They were not officious or obnoxious; they were good communicators. They were fair. As a result, they made many a case where a more officious officer would have failed. Notable among them are Steve Reynolds, Jim Low, Jane Scheid, Gary Folger, Curt Youngren, Ted Ruddell, Joe Abrams, Dick Fueling, Bob Boutang, Larry Hensley, Dave Lorring and Dick Hemmen. OK, I'll admit that Bob Boutang was a bit squirrelly, but I loved him like a brother, even when we rented a house together in Anchorage during the time we each held down staff jobs in our respective State Directors offices. It was probably not that we had earned our way close to the top of the food-chain, but rather our Directors wanted us where they could keep a close eye on us!

I would still be fumbling around in the snow trying to figure out how to trap in the subarctic of Interior Alaska, if it were not for the tutelage of some accomplished trappers who were willing to share important things that they knew. Those that come to mind include Norm Phillips Sr., Ron Long, Terry Johnson, Danny Grangaard, Jon Gleason, Bear Wyse, Ken Deardorff (whose lovely daughter Susan married my son Jason) and of particular note, Ben Hopson Jr. of Annaktuvuk Pass. Ben knew where a wolf was going to step, before the wolf did!

I would be remiss if I did not mention a number of ADFG wildlife biologists who were consistently loyal to the science of predator-prey relationships in the face of relentless adverse political pressure and vitriol. Among those for whom I harbor a ton of respect in this regard are: Bud Burris, Dick Bishop, Mel Bucholtz, Larry Jennings, Dan Timm, Jim Faro, Ted Spraker, Patrick Valkenberg, Bill Gasaway, Dave Kelleyhouse, Bob Stephenson, Wayne Heimer, Ron Boertje, Roger Seavoy, Toby Boudreau, Don Young (the biologist one...), Tony Hollis, Jack Whitman, Tom Seaton, Mark Keech, and Cathie Harms. Again, I probably missed some; mea culpa.

I am fond of saying that I while I harbor much respect for my elders, the immediate and increasing problem is that my elders comprise a shrinking gene pool!

I am lucky to have my hunting travel arrangements coordinated by Karen Gordon of Fairbanks. "Gordie" worked with me at Alaska Division of Forestry and in fact did my travel. "Gordie" also proofread the manuscript and schooled me on the use of the Oxford comma! We both retired, but she still volunteers to keep my travel plans all nice and coordinated.

I am indebted for coaching from other friends who had dived into becoming authors before I. Sharon McLeod wrote her book about guiding long before I wrote mine. She titled it Walk Softly With Me; it should have been called "She Who Walks With Stupid Men." Randy Zarnke, my friend of many years has written some fascinating books. At times I was recruited to help with proofreading or editing chores so I had a rudimentary idea of what was involved. Randy was very patient in relating over and over the rest of the things that he had a good grasp of and I had never considered. Bill Brophy was a long-suffering proofreader who converted my rambling into actual sentences with grammar, punctuation and everything! Ryan Ragan worked hard to teach me the meaning of the word "format;" I am grateful and so I'm sure, are my readers.

My friend Ken Coe, a retired smokejumper who shares a penchant for colorful story-telling and who speaks and writes clearly, also reviewed the manuscript in its final stages.

Table of Contents

Glossary 11

Chapter 1: Me Caribou is on Fire! 16
Chapter 2: Beaver Creek Moose Hunt 24
Chapter 3: What's Left to Hunt? 32
Chapter 4: "Plains Game" at 7,000 Feet 44
Chapter 5: How to Draw an Alaskan Bison Permit 54
Chapter 6: Fotch Creek Bull 62
Chapter 7: Emma Lee's Kudu 68
Chapter 8: Arctic Death March 78
Chapter 9: Mentors 84
Chapter 10: So You Want to be a Guide? 94
Chapter 11: Cowboy Mexico 100
Chapter 12: The Aussies Try to Blow Me Up 112
Chapter 13: My Name is Tumkulu 126
Chapter 14: Tanana Flats Death March 142
Chapter 15: Hunting in the Land of Hobbits 150
Chapter 16: Hunting Dinosaurs 162
Chapter 17: Mekoryuk Muskox 170
Chapter 18: Rams on Ice 180
Chapter 19: Kodiak: One Shot Per Day 194
Chapter 20: The Big Chill 206
Chapter 21: Pitch 'Till You Win! 222
Chapter 22: Walking Wounded at North Fork 236
Chapter 23: Buckaroo Stage 246

Foreword - *By Pete Buist*

Youngsters who live in small flat Eastern states yet are enamored of the great outdoors only have muskrats, 'coons, panfish, cottontails, a few reptiles and the odd unlucky "tweety-bird" to intrigue them and tide them over. Most get to hunt big game only in the form of whitetail or maybe mule deer. In my youth, on the pages of Outdoor Life, Field & Stream, and Sports Afield were the foundation of my dreams. Those publications were the source of the awesome adventures I craved. My Mom and her side of the family had always been shooters and hunters, but it was the fault of those magazines that I became hooked. On those glossy pages, two places in the world were featured prominently as THE places big game hunters visited. They beckoned to me irresistibly.

The first place was Alaska. Alaska may be on the same continent as my hometown in New Jersey, but that is about the only similarity. Alaska has an abundance of huge animals, vast wilderness, spectacular mountains, expansive tundra, and magnificent glaciers. It was and is a hunter's paradise. In my youth, Alaska was a territory, not a state. Statehood would come later. Nonetheless, Alaska beckoned to me. The other place of course, was Africa. Land of dangerous game, the Big Five, oodles of antelope and exotic safari tent camps.

I was lucky. At age 23, thanks to being drafted into the U.S. Army, I was stationed in Alaska. Clearly the Army, not to mention my friends and neighbors at the local Draft Board, felt strongly they could not be certain that the world would be safe for democracy unless I took up arms and participated. Under some duress, I escaped from New Jersey and did my military service in Alaska. An Alaska assignment was considerably more conducive to longevity for an Infantryman at that point in history! I proudly served in the 172nd Arctic Light Infantry Brigade, got out of the service and stayed in Fairbanks, Alaska. Before I turned 30, I had successfully hunted and taken, among other game, moose, caribou, black bear, Dall sheep

and brown bear. I had been elected as a Director of the Alaska Trappers Association. I had an Alaska Assistant Guide's license in my pocket. I had survived a slight mishap in a bush aircraft. I had had close calls with bears, but had not actually been munched on.

I slowed down a bit and methodically hunted most of Alaska's 13 species of big game. Later I earned a Registered Guide Outfitter license and ultimately was awarded Alaska Master Guide license #79. As I somehow survived to the half-century mark of my life, the "other" place of my childhood dreams hauntingly distracted me. Africa was calling to me!

In 2004, I retired from the Alaska Department of Natural Resources, Division of Forestry where I administered timber sales in winter and managed wildland fires in summer. My goal was to do a bunch of things such as hunting in Africa before my body went "over warranty." Since then, I've been on seven African safaris! And I've hunted in a lot of other countries as well. I've made a lot of great friends and had a lot of fantastic adventures.

Robert Ruark, author and adventurer, wrote that a hunt is comprised of three parts. I concur. The first is the anticipation, including planning. Next is the hunting adventure itself. Finally, there are the memories. The details may fade with time and age, but the memories are precious. Since the early 1970s when I first arrived in Alaska, I tried to keep journals and diaries of my hunting adventures and expeditions. I enjoy documenting the stories and sharing them. More than a few people suggested that I write a book. What better way than to share and relay the content of some of my more memorable adventures? My journals describe some hunts where the game we sought were harvested and others document adventures where no wild animals were harmed.

There are winter trips and there are summer trips. There are trips punctuated by accidents and surprises. The common thread among these adventures is that I lived through it all and I had FUN while hunting.

As long as I am able AND having fun, I will continue to go hunting. With luck, I'll be able to continue sharing these fond memories. Maybe they will inspire some budding hunters to make the effort to live their dreams. Maybe they will help others avoid some of the near-death episodes I experienced and luckily survived! At the very least, I hope they will provide enjoyment for those who read on.

Meanwhile, allow me to share some fun and exciting expeditions with you. Let's go hunting. There is a world of adventure out there!

Glossary

<u>ADF&G</u> – Alaska Department of Fish and Game.

<u>ATCO Unit</u> – Modular office or living space units. They were common during pipeline construction. Many from that era are still parked here and there in rural Alaska.

<u>Baculum</u> – Bone found in the penis of many placental mammals, including bears.

<u>Bakkie</u> – South African term for a pickup truck. Generally open, but may feature a rack. Distinct from a large truck, van or SUV.

<u>Belly pod</u> – A sealed baggage compartment attached to the underside of bush aircraft.

<u>BOQ</u> – Military term for Bachelor Officers' Quarters.

<u>Braai</u> – Afrikaans for barbeque or roast. Sometimes also the word used for the "fire pit" or clay oven on/in which the cooking occurs.

<u>Bush fires</u> – Australian term for what we call in the U.S. wildland, or "forest" fires.

<u>Boone and Crockett Club (B&C)</u> – A club founded in 1887 by, among others, Theodore Roosevelt, to promote conservation, wildlife management and ethical hunting via "Fair Chase." The club maintain the Records of North American Big Game.

<u>Cache</u> – A storage structure in rural Alaska. Normally elevated and reached only with a removable ladder. Used to keep marauding bears and other creatures out of one's supplies. Also used at remote cabins to store a tent and stove as a hedge against being burned out of one's cabin.

<u>Catch (an animal)</u> – Common terminology in the Athabascan culture for the "taking" or "shooting" of a game or fur animal or fish.

<u>Chicos</u> – New Mexican term for dried sweet corn. Commonly rehydrated and served with an assortment of meats.

CITES – Convention on International Trade in Endangered Species. An international agreement between governments. Originally designed as a legitimate conservation tool for insuring that trade in certain species of plants and animals does not threaten the survival of the species, it has unfortunately evolved into a political process that pits conservation agencies against non-governmental organizations with ulterior political motives.

Coot – An articulated 4 wheel drive all-terrain-vehicle. Most were produced in the 1960s and 70s. Production ceased soon thereafter with the advent of 3- and then 4-wheelers. The fact that many of them still are in use in Alaska is a tribute to their hardiness.

Cuppa – Australian term for a refreshment break or merely a cup of tea or coffee.

Dagga Boy – African term for Cape Buffalo, normally older, non-breeding bulls who stay solitary, or in bachelor groups rather than mingling with the breeding herds. Dagga is the Shona word for mud. The older bulls often wallow in mud and coat their skin with it.

Eastern Cape Province – of South Africa. The southeastern most province of the Republic of South Africa, bordering on the Indian Ocean. Capital is Bhisho. The largest cities are Port Elizabeth and East London.

Fallow Deer – (Dama dama) A species of ruminant mammal belonging to the family Cervidae. Native to Europe, but has been widely introduced around the world, including South Africa, Australia and the U.S.

Fold-A-Sled – A metal freight sled designed to be pulled behind a snowmobile. Popular in the 1970's before lighter plastic sleds were invented. It was designed to fold down fairly flat for easier transport, although Alaska trappers found that if it was left in the manufactured state to allow folding, it soon lost a lost of structural strength and was prone to breakage. This problem helped it to earn the nickname "Weld-A-Sled!"

Game Management Unit (GMU) – Alaska encompasses 26 units and subunits for the purpose of game management.

Gang Chain – Alaska mushing term for a chain or cable with short lines spaced along it for securing a dog team once it is out of harness.

Guide/Outfitter – Statutorily defined term in Alaska law for individuals licensed to provide "big game commercial services" (guiding) in the field.

Hind – European term for females of the deer species.

In the salt – African expression for game and hunting trophies that have been prepared for storage or transport. When you have taken your trophy and it is back at the camp or lodge, it is said to be "in the salt."

Inja – Xhosa for domestic dog.

Koek – Afrikaans for cake.

Kopje – Afrikaans for small hill in a generally flat area.

Kraal – Afrikaans or Dutch, for corral for livestock. Sometimes also used to describe a group of living quarters, generally enclosed by a fence or wall.

Lapa – A South African structure, normally without walls and often with a thatched roof. The word is a Sotho and Tswana word that actually means "home." It usually refers to a communal dining and lounging area, separate from sleeping quarters.

Limpopo Province – Formerly the Northern Province of South Africa. It is the northernmost province and is named for the Limpopo River which forms the western and northern borders (with Zimbabwe, Botswana and Mozambique.) The capital is Polokwane.

Loppers – South African term for brush cutters.

Metebele – A subgroup of the Zulu people driven out of the Transvaal and into Rhodesia by the Boers in 1837. Now more commonly known as Ndebele, which is also a dialect related to the Zulu language.

Mopane – (Colophospermum mopane), commonly called balsam tree, butterfly tree or turpentine tree. In the legume family, it grows in hot, dry low-lying areas of northern parts of southern Africa.

Ommingmak – Eskimo term for musk ox.

Pan – shallow African water hole. In RSA, often with a solar-powered pump in a bore hole.

Panga – African version of a machete or brush hook tool.

Pap – Traditional porridge made from maize-meal (coarsely ground maize) and a staple food of the people of southern Africa. The Afrikaans word pap is from the Dutch and means "porridge."

PH – Common abbreviation for Professional Hunter. A licensing category of hunting guide in Africa.

Potjie – Afrikaans for stew

Qamutiik – Eskimo word for long wooden sled, pulled by snow machines or sled dogs. Alternatively komatuk

Rondavel – An African hut that is ordinarily in the shape of a cone on cylinder or cone on drum. From the Afrikaans word rondawel. Normally fashioned from native materials with a wood frame and thatched roof. May also have a stone or masonry wall.

Rusa Deer – (Rusa timorensis) More properly Javan rusa. A deer native to Indonesia and East Timor. Populations have been introduced in a wide variety of locations across the Southern hemisphere including Australia and New Zealand.

Safari Club International (SCI) – Similar to B&C, but worldwide in scope. SCI's goal is to protect the freedom to hunt and promote wildlife conservation around the world.

Sambar Deer – (Rusa unicolor) A large (400 pound) Asian deer native to the Indian subcontinent, South China and Southeast Asia that has been introduced to several places, including Australia and the west coast of the U.S.

Same-day-airborne – Refers to a law in Alaska where hunters generally cannot hunt or take big game the same day they have flown in small aircraft. The law is designed to stymie the old practice of spotting game animals from the air, then landing and shooting them.

Scooter Bull (Kudu) - Refers to young kudu whose horns make only a twist and a half. The points are directed outward, like the handles on a motor scooter.

Shooting Sticks – Tripod of poles, popularly carried by African PH's and set up for a client to rest a rifle on for a shooting rest.

Skookum - Alaska slang for something that is efficient and/or useful.

Snow machine - Alaskan term for snowmobile. Also sometimes referred to as a "snow go."

Spike Camp – Or fly camp. A temporary camp, away from a base camp or lodge, which allows a hunter to stay (eat and sleep) closer to the game being pursued. By staying in spike camp (or "spiking out,") a hunter can concentrate on hunting an area and not have to worry about getting back and forth to the hunting area each day.

Squib Load – Refers to hand-loaded ammunition that contains an inadequate amount of gun powder. Occasionally a squib load, when fired, propels a bullet only partway up the barrel. This essentially means that the firearm now has a blocked muzzle and is very dangerous.

Stag – European term for males of the deer species.

Torch – British/Australian term for flashlight.

Trackers – General African term for staff who accompany a PH and his client(s) in the field on a hunt.

Transporter - Statutorily defined term in Alaska. Licensed Transporters furnish various forms of transportation for hunters, but in general, cannot provide other "big game hunting commercial services" in the field.

Ute – Australian term for a pickup truck.

Wet Lock Boxes – Waxed corrugated containers designed specifically for shipping fresh fish and meat. They come in a two-piece telescoping gusseted design to ensure water/blood resistance and insulation when used with a poly liner.

Xhosa – The Bantu language of the Xhosa, a South African people traditionally living in the Eastern Cape Province. They are the second largest ethnic group in South Africa, after the Zulus. It is one of the official languages of South Africa and is spoken by over 7 million people.

Zim – South African slang for the country of Zimbabwe.

Chapter 1

Me Caribou Is On Fire!

Even in my later years guiding in Alaska, I never had a huge operation. It had always been "Mom and Pop" and would remain that way. Part of the fun of guiding is the sharing of what we have and love, with others. When clients from foreign countries had similar adventures in places I wished to explore and experience, we often worked out "swap hunts." As a licensed guide, I had the option of trading hunts to hunters, particularly other guides that the average Alaskan hunter would not have been able to legally do. I could take out-of-state and even out-of-country hunters without running afoul of Alaska law. This is how I came to trade several hunts with Australian "mates," Dan Field and Jeff Garrad of Narooma, NSW and G. Ross Ferguson of Wangaratta, VIC, Australia.

Dan had originally set up a swap hunt with another Fairbanks resident. In fact, that hunter had already been to Australia and hunted pigs and goats with Dan. Now Dan and Ross were headed to Alaska to get their swap hunt for moose and caribou. A month or two before they were to arrive, their Fairbanks contact realized that non-resident ALIEN hunters must, by law, contract with a Registered or Master Guide. Another U.S. hunter could have hunted moose and caribou completely unguided. Not so with the foreigners. The Fairbanks hunter was begging me to take the Aussies. I worked out my own swap hunt deal with them, ultimately agreeing to take them for moose, caribou and grizzly. We were off to the races. It was to be the start of a fantastic friendship and some exciting adventures to be sure. Some of the adventures nearly killed us!

In late August of 1996 we headed up the Dalton Highway (Haul Road.) My partner Dave was already in Coldfoot. He had been flying some buddies into his cabin for a moose hunt on the Middle Fork of the Chandalar River. The Haul Road was in terrible shape with rain and mud. It took a solid seven hours to drive from Fairbanks to Coldfoot. I could see 50Q, Dave's Super Cub, parked by our trailer on the old airstrip behind the main buildings at Coldfoot. We had a grand round of introductions. After a bit Coldfoot

Me Caribou Is On Fire!

You rarely see a roadsign like this one just north of Livengood, AK on the Haul Road.

manager, Troy, drove in bearing dire news of a weather warning. Blizzard conditions forecast for the entire Eastern Brooks Range. A lot of people were hung up at Coldfoot, trying to disperse to the North Slope. We dug in. Troy happily began envisioning all the money he was about to make.

We briefly had the Coldfoot Saloon to ourselves. After a while, a busload of tourists descended upon Coldfoot and the place filled up. Troy comped us a couple rooms in the Inn; Dan and I took one, Dave and Ross the other. Assistant Guide Glen and packer Gary bunked in our travel trailer by the airstrip. We liked the Coldfoot Inn better for now. Glen, Gary and I abstained from the booze; the rest of the crew... well, did NOT! Aussies and beer are a classic combo. Said combo was on full display. About 9:30 pm Gary, Glen, Dan and I went over to the café to eat supper. Ross and Dave said they would be along soon, but it was not to be! When we finished up, we trudged back over to the Inn; Dave and Ross were still there. They ate later,

Me Caribou Is On Fire!

Dan Field modeling the hung-over look at Coldfoot, post hunt.

and Dave mentioned that Ross fell asleep during dinner. Ross's description? "At least I didn't dip me nose in me tea.....!"

The next morning there were some hangover issues, but nothing serious. Overnight there had also been a fair amount of snow accumulating on the lower peaks, but I decided to try and make Happy Valley, a distance of about 140 miles, before the brunt of the storm hit. While the concept was sound, this turned out to be only a marginally thought out decision. I hooked the trailer to my half ton Ford and pulled out onto the Haul Road. Our other vehicles fell in behind me.

Atigun Pass is an impressive but fairly inhospitable piece of northern Alaska real estate. I never cross it without thinking in awe of the engineering and construction effort it took to get a road and a pipeline over it in the early 1970's. We wound our way up the south side and then down the north side. A road grader had been over the top and pushed most of the new snow aside,

Me Caribou Is On Fire!

but the drifts were quickly blowing back in. We got video of fox, ptarmigan and sheep and made it to Happy Valley in a shade under six hours of horrible road conditions. More to the point, we made it in one piece with no wrecks.

At Happy Valley we pulled in beside Registered Guide/Outfitter Len Mackler and his dad, Ray. We got the trailer arranged to serve as a kitchen and dining area, and set up a couple of wall tents for gear storage and sleeping.

The following morning we said goodbye to Len and Ray. They would be driving south, so we asked them to check on Dave and Ross at Coldfoot and relay to them what the weather was like in the pass. Atigun Pass is not only inhospitable to ground vehicles, it is notoriously dangerous for small planes. It is roughly S-shaped. Unless you KNOW you have clear weather on both sides, you should not attempt to fly though. By the time you reach the middle of the "S" and find the other half weathered in, you do not have enough room between the canyon walls to turn the aircraft around. The pass is littered with airplane parts commemorating the downfall of pilots who have failed to heed this advice. As it turned out, Len, a pilot himself, decided that the pass was un-flyable. He advised Dave to fly through the much wider Anaktuvuk Pass, further to the west. Several pilots flew in during the day (via Anaktuvuk) to stage out of Happy Valley for caribou hunting to the northeast around Kavik. We were sure Dave would make it to Happy Valley this day. Ultimately he did.

Dan Field is nothing if not innovative. An Aussie who is both bored and innovative is a joy to be around, particularly so if beer is part of the equation. We had just finished breakfast when Dan invented "catch and release fishing for ground squirrels." It's a simple process really; it's a wonder no one had already thought of it. You take a small fishing lure, file down the barbs, bait it with peanut butter, and cast to one of the many ground squirrels that call the Happy Valley airstrip home. Since squirrels love peanut butter, it didn't take long to hook up and "play" said squirrel like a fish. The best part turned out to be watching the Aussie put on heavy leather gloves and do what it took to release the angry squirrel.

Later Dan regaled us with a tale of how he discovered and shot an "intruder" in his house. Seems it was Christmas Eve and after the party had calmed and

Me Caribou Is On Fire!

the family had gone to bed, Dan thought he heard a noise. Arming himself and creeping toward the lounge room where the presents were around the tree, he spotted the culprit and opened fire. What he ended up "killing" was actually a cardboard cutout of Batman that had been left standing by the tree for one of the kids!

As entertaining as it is having Dan around camp, it seemed like a good idea to get him out on the tundra to actually do some hunting. Once Dave and Ross flew in, it was high time to get the hunters out to spike camp. But due to oncoming nightfall, that task would have to wait for the next morning.

Dan was a little difficult to call for breakfast. As he explained it "I couldn't hear you calling. I had martens in me ears!" Dan had grown quite attached to my marten trapper hat and had been wearing it 24 hours a day, including in his sleeping bag. It was completing his Aussie "mountain man" look quite nicely. I had a feeling that he was WAY attached to it and I was right. When he left to return to Australia "Martin" was in his duffel bag.

By lunchtime Dave and Glen had scouted out a nice place to camp and hunt, about 50 miles west of Happy Valley, down on the lower Itkillik River. There were plenty of caribou around and sign of grizzlies, so it would make a great home for at least a few days. Dave dropped off Glen off to set up camp and flew back to get Dan.

Dave and I checked on them the next afternoon. Dan had his grizzly and was now actively looking for a nice caribou bull. We decided that since things were going fairly smoothly, I should go out with Ross and see if I could help find him a grizzly. Rather than going all the way back to Happy Valley, Dave dropped me at an old oil exploration site and went back for Ross. Just at dark, Dave flew in with Ross and a bit of food. Ross and I had enough light to do a quick recon of the area and were pleased to observe both caribou and a pretty fair, light-colored bear, all within a couple miles of our camp. While we never saw that particular grizzly again, just a couple of days later, Ross and I stalked up on and took a fabulous large dark-colored bear that ended up making the B&C record book!

The following day, Dave picked us up midday and moved us back to Happy Valley. Afterward he checked on Glen and Dan. They were "done" but the adventure had not been without its moments. From their perch on a cut-bank

Me Caribou Is On Fire!

Dan Field shows a North Slope grizzly he took with his black powder rifle.

above the river, they had spotted a small group of caribou bulls on a gravel bar less than a mile away. One of the bulls was spectacular and Dan headed out to stalk it. Using his paper-cartridge black powder 40-90 with open sights, he had to get fairly close. He ended up closing the distance to within 120 yards. At Dan's shot the bull dropped. Dan turned to look at Glen and celebrate. When he looked back, the bull was back up and staggering about the gravel bar. Dan reloaded and began approaching the animal, getting to within just a few yards. Now that caribou bull could have staggered in any of 360 degrees, but he somehow chose to proceed directly toward Dan. Or, as Dan describes it: "He charged me...!" The caribou lurched ever closer. Finally he was just a gun barrel length from Dan. Dan fired from the hip, with the muzzle touching the long white hairs on the caribou's neck. The projectile entered near the top of the bull's sternum. It exited somewhere south of his nether region and drove into the rocks and gravel. Down went the bull, but also down went Dan, knocked unconscious by a piece of river rock that flew into the air and came down, hitting him squarely on the top of

Me Caribou Is On Fire!

Ross Ferguson with his dark colored North Slope grizzly bear.

his head. "Martin the Trapper Hat" had been left behind for the stalk. Said hat would have been better worn as protective head gear.

Dan awakened from all this noise and drama in a sitting position. His rifle was across his lap and there was a dead caribou lying with its head on his feet! Blood was streaming down Dan's face from where the piece of rock had cut him. Seeing all this but not knowing exactly what had transpired, Glen ran up and asked Dan what had happened. Dan's reply? "I think I shot me-self!" As one might imagine, this thought was somewhat disconcerting to Glen, who naturally was concerned for his client's general well-being. He began explaining to Dan that it might be several days until Dave could fly back over to check on them. In the midst of his rant, he notices that Dan is kicking the dead caribou. "Now what the hell is going on?" Dan utters the iconic phrase "Me caribou is on FIRE...!" As it turned out, the burning black powder that exited the muzzle of the 45-90 had in fact ignited the hair on the caribou's neck! Dan, by this time convinced that he had NOT shot himself, but rather had been cut by fragments of river rock, was merely dutifully trying to extinguish the burning caribou hair!

Me Caribou Is On Fire!

Dan Field and his caribou ... after the fire was extinguished.

Dave was able to fly the next day and pick up the "wounded" hunter and agitated assistant guide, and fly them back to Happy Valley. In a couple of days we broke camp and drove back south. We had to listen to Dan's rendition of the burning caribou adventure, ad nauseam, all the way down the Dalton Highway back to Fairbanks!

Chapter 2

Beaver Creek Moose Hunt

My friend Fred, who has a nice cabin north of Fairbanks, at the Big Bend on Beaver Creek, in GMU 25, contacted me one summer. He had bought the cabin and piece of land from an old trapper, Herman Bucholtz, in the 1970's. "Herman the German" as he was known, had built that main cabin and trapped from it for many years. It had primarily been a good marten line. Herman trapped it all on foot, accompanied only by a companion dog; he never had any interest in using a dog team. He was a tough old guy, but when he was diagnosed with Parkinson's Disease, he had to give up trapping. He died soon thereafter. Fred had done some flying work for Herman. His relationship with the family was such that he was able to buy the cabin and 40 acres from Herman's estate. Herman had another 10 or 15 smaller "line cabins" but they were not nearly as spacious and nice as the main cabin at Big Bend.

Herman's nephew, Hans Goetz, lived in Germany. He had discussed with Fred the possibility of coming back to Alaska for a moose hunt. Hans had in the past visited his uncle and stayed at the cabin. He wished to return and bring his daughter Anja to see the beautiful wild country in the Big Bend area. Fred knew that he could not legally allow Hans or Anja to hunt moose without having a Registered or Master Guide along to make the adventure legal and thus he called me. I was happy to oblige.

I already had some on-the-ground experience and history in the area. Herman "sold" his trapline in a couple of pieces. The Alpha Trail and the Beaver Creek Winter Trail (from 23 Mile Elliott Hwy) had originally been purchased by one Fairbanks trapper. The downstream portion, including the Summer Trail (in from 28 Mile, Wickersham Dome) Beaver Creek itself and Fossil Creek, had been sold to two other trappers, Leroy and Jerry. They only trapped it one season before Jerry died in a snowmachine accident and Leroy moved to Kantishna to trap and work as a lodge winter-caretaker. Leroy allowed Gary Thompson and me to trap their portion and we did so for a couple of years. It WAS a terrific marten line. In our best season, the

Beaver Creek Moose Hunt

Herman's cabin at Big Bend on Beaver Creek, north of Fairbanks.

winter of 1976-77, Gary and I took 213 marten and five wolverine. This hunt would be my "homecoming to Beaver Creek."

Fred flies me over from Fairbanks on the night before the hunt. Hans is a corrections officer at a German youth penal facility. He retired from military service where he was a Master Sergeant and tank commander. He speaks English "wery vell!" His daughter Anja, is 17 and does not speak English nearly as well as Hans. I speak no German at all and depend on Hans to translate as we discuss plans. Anja is to have first crack at any moose we might spot.

Anja has passed the rigid education and testing program required to hunt in Germany and is an accomplished hunter in her own right. She has hunted and taken roe buck in Germany. The German hunting tests cover a plethora of aspects of hunting including practical applications such as shooting prowess. She has long dreamed of hunting Alaska moose in the footsteps of her great uncle Herman. Our objective is to assist her in living that dream.

25

Beaver Creek Moose Hunt

The "hochsitz" (high seat) built by Herman as a perch from which to glass for moose.

Although it is September 4 and normally would be crisp fall weather, unfortunately it is quite warm. We didn't even bank the stove last night. We just fell into our bunks and corked off. One more thing I will say about Herman... building a large main cabin that sports four queen-sized bunks and mattresses was a capital idea! Fred, Hans, Anja and I make good use of those big bunks. Fred's two sons, John and Bill, have taken up outlying quarters in a tent alongside the airstrip. I sleep soundly until about 3 am when I need to step outside. I come dangerously close to peeing prematurely when I nearly trip over a marauding porcupine perched on the top step of the porch in the dark.

There are no snorers in the cabin crew (unless it was me and I was blissfully unaware.) Anja talks in her sleep a bit, but since the conversation was all in German, she revealed her secrets only to Hans.

Beaver Creek Moose Hunt

Opening day is damp and way too warm for what we normally consider good fall moose hunting weather. The fog is down on the deck and visibility nil. No sense in going out to glass. By 11 am the fog begins to lift. We take the time to get Anja sighted in at the range. By mid-afternoon we are hunting.

As this is a proper German hunting cabin, it comes with a "hoch sitz" or high seat. Herman built this one years ago as a glassing platform from which to watch for moose out to the west and up onto the ridge separating the Beaver Creek drainage from that of the East Fork of the Tolovana River. This particular High Seat also has an outhouse at the bottom, which we refer to lovingly as the "Low Seat." Also at ground level is Hannah, Fred's German Shorthaired Pointer who is seriously working on running every single red squirrel out of GMU 25C. From on high, I spot no moose. Cows, bulls or otherwise.

Around 6 pm I begin preparing a 12 pound silver salmon for dinner. I rinse and season the fish and stuff it into the oven with some foil-wrapped potatoes to bake. The hunters all flock back into camp around 7 pm, but the only living things spotted all day have been a couple of other moose hunters floating Beaver Creek.

There is still plenty of light left, so after supper the whole crew heads back out to a good glassing point to see what the twilight may reveal. By 10 pm it is black dark; we stumble back into the cheery, warm cabin.

The following day dawns as foggy as the first. Discouraged, Hans and Anja decide to try fishing Beaver Creek for some Arctic Grayling before breakfast. After a couple hours we hear a single-engine plane overhead, which leads us to believe the fog is concentrated in low areas and we might expect a break in it soon. An hour later, ADFG biologist Toby B walks up to the cabin to visit. The plane we had heard was Sandy, a Tamarack Air pilot, bringing Toby and his buddy for a float trip and moose hunt on Beaver Creek. They are planning on floating from here at Big Bend, and taking out at the airstrip at the old Miller homestead at the mouth of Victoria Creek.

Anja and John glass from the bluff for quite a while, eventually spotting a little paddle horn bull and a bigger moose back on our side. They pile off the

Beaver Creek Moose Hunt

bluff and make their way over to the opening where the moose was originally spotted. After a bit, both bulls reappear.

Anja is able to take an offhand (standing) shot at about 110 yards, downing the bigger bull. Proving the legendary toughness of Alces alces the bull jumps back up and into the pond. But thankfully, he then charges all the way to the opposite side and out onto terra firma. Anja puts him down for good.

Darkness is coming on fast. We accomplish the gutting of the bull essentially by Braille! The float back down to the cabin in just about pitch dark is an exciting naval maneuver!

The reheated dinner is gratefully received by our young Diana. Today she has lived her dream. She has taken a gorgeous Alaskan bull moose with a 58 inch spread and memories for a lifetime. We pile into our bunks around 1 am. Unbeknown to us, the "adventure" is far from over!

The next day, September 7 is clear and cool. Better moose hunting weather. But instead of moose hunting, we have work to do. We plan to do all the meat recovery by landing the Bellanca Scout on the big gravel bar, then using the canoe and river boat to move the meat out to the main river. It is a chore locating the kill; the area looks very different in the daylight. We finally spot the activity of the Gray Jay "camp robber" birds that are frolicking and feasting on the moose guts. There is Anja's bull.

After we take plenty of photos and video, the hard work begins. Even with some brushing and trail establishment, this chore is not for the weak of heart or weak of back. Utilizing all the help we have available, we get Anja's bull caped, skinned, quartered and ready to pack out to the river in under two hours. Fred takes some of the meat in the canoe and paddles out to the gravel bar. He loads it into the Scout and takes off for Fairbanks. We head back to our temporary landing in the slough to bring out the rest.

At the landing we load some meat into the canoe with John and Bill. The rest goes into the small river boat. It doesn't look like much of a load, so we add the antlers, the packs, Anja, Hans, John, and me. To prove that there is still free-board, Hannah the dog jumps in and perches on the pile of gear. The fact that we are severely overloaded becomes clear as soon as we leave the sheltered water of the slough and into the current of Beaver Creek!

Beaver Creek Moose Hunt

Anja with her trophy bull moose from Big Bend.

Oops! Water begins pouring in over the gunwale and filling up the boat. As a highly trained and observant Master Guide, I immediately take this to be a fairly bad sign. My reaction is to holler "Oh shit….!" as loudly as I can. Hans and Anja gave a similar "Oh sheis!" Over we go. The water in an Alaskan river in September is quite cold, so it turns out!

Thankfully, Hans grabbed both rifles and bailed over the left gunwale, the shoreward side. He is close enough to the gravel bar to stand. I leaped out the opposite side, and for some reason think to grab the bow line as I go. The water is up to my chest and colder than a Greenpeacer's heart.

Anja has made a valiant effort to save her moose antlers, but to no avail. She has reached the gravel bar and is shivering while mumbling "Mein elche, mein elche…" On the bottom of Beaver Creek are five white linen bags of moose meat (close to 250 pounds) and the antlers of her bull. Our trusty little boat, in its soggy condition, probably weighs more than a ton and will not float on a bet! It rests solidly on the bottom in shallow water.

Beaver Creek Moose Hunt

Moose hunters heading out to look for moose along Beaver Creek.

Paddles, coats, bags, PFD's and assorted light items are floating merrily away down Beaver Creek. I am able to intercept most of them and get them flung up on the bank for later collection. After 20 minutes of bailing, the boat had lightened and risen far enough to be flipped over and drained of its remaining river water. The next major task will be the recovery of the moose meat, which I should add, is by this time nicely chilled from its half hour spent on the bottom of the icy river.

Since I'm completely soaked anyway, I strip down to my long johns and dive for the bags of meat. Being light-colored, they are relatively easy to locate. Anja's antlers however, are nowhere to be seen. I'm very afraid the current has swept them much further downstream. Locating them is thus far more problematic, simply because their color blends with the color of the gravel on the river bed. As luck would have it, this particular set of antlers are still stained a little bit red from the alders on which the bull had scraped off the velvet. I climb a high bank and use my binoculars to scan the river bed. Finally I spot a reddish glow at the bottom of a deep pool among the cobbles

Beaver Creek Moose Hunt

Mountains that form the route of Beaver Creek around the Big Bend.

and mud. One more deep dive into the freezing water for me and finally Anja's "elche" is rescued from its watery resting place.

By this time we are all borderline hypothermic, but after a bit I come to realize one of the rescued bags is my "dry bag." Note to self: I should have had my camera in the dry bag instead of in my shirt pocket! But inside the bag is a dry winter jacket. Since I am a naturally chivalrous sort of fellow, I haul out the jacket and offer it to Anja. Bless her heart, the uninhibited fraulein peels off her wet gear and puts on the dry jacket. Her dad is still shivering, but the rest of us are warm once more!

A few hours later we are all back at the cabin. Fred has flown the remainder of the meat back to Fairbanks. We are enjoying fresh moose tenderloin from the grill, along with steamed rice and broccoli. We are exhausted, but grateful for good food, a warm cabin, and no one the worse for wear.

International friendships, crisp fall weather, the smell of wood smoke, freshly split spruce bolts, boiled coffee, bacon, gun oil, and pipe tobacco all combine to make this the exquisitely high-quality experience that those unfortunate people who do not hunt are not privileged to know! We who hunt are indeed blessed.

Chapter 3

What's Left to Hunt?

By the time I began my seventh decade of life, I had been to Africa. In fact, I had been on several safaris and had taken a nice array of "plains game." But I hearkened back to my formative years. Back to the pages of Outdoor Life, Field & Stream, and Sports Afield. I had not hunted dangerous game in Africa. The Big Five! I was retired and could finally afford (barely) a dangerous game hunt. Maybe not the really expensive stuff like lion or elephant. But Cape buffalo seemed to be a reasonable way to scratch this particular itch. I booked a buffalo hunt for April of 2016!

Emma Lee and I had spent the previous month in the Eastern Cape Province of South Africa. We had married the previous spring, but this was our long-anticipated "hunting-moon!" The first phase of the trip had been with my PH friend John "Madala" Barnes chasing *n'gulubi* (bush pig) down on the coast of the Indian Ocean at St. Francis Bay. From the coast we headed inland for a mixed bag hunt for plains game with my boys Jason and Peyton and their wives, in the Grahamstown, RSA area. All the hunters in the family had taken an array of plains game. I could have been satisfied to head back to Alaska. But knowing I was going to be going to the Eastern Cape anyway, I had added a booking for a Cape buffalo hunt up on the Botswana border as the grand finale of our African trip. International airplane tickets don't come cheap. Why not make THREE safaris out of one round-trip to Africa?

From Port Elizabeth where we left John, we flew back to Johannesburg on South African Airways and stayed overnight at Afton House. We were met the following morning by our PH Steyn Potgieter of Polokwane for the buffalo hunt. We jumped into Steyn's already fairly well loaded bakkie and headed for his family property and lodge on the Limpopo River, a delightful retreat called Sekombo River Lodge.

It is 9 am when we leave Afton House and nearly 4 pm when we arrive at Sekombo Lodge. Those seven hours allow for plenty of rookie questions from me and trust me I asked them. Sekombo is a 1,250 hectare farm that

What's Left to Hunt?

African tracker using his smart phone.

has been almost completely converted for hunting use. There are a few sheep and cattle as well as vegetable gardens and some hay fields. But the property has been almost completely repurposed for hunting and eco-tourism. It ranges from thick mopane and thorn brush to towering shade trees along the "great greasy green Limpopo" itself. At this time of year, the river is fairly low, with deep tranquil pools among the flat river rocks. By walking up or down river just 100 yards, you can normally walk easily across… and be in Botswana! There is an excuse for a fence around the exterior of the property. It is probably pretty effective at keeping game such as impala, nyala, waterbuck, wildebeest, zebra and blesbok convinced to stay on the property. The chances of it holding in the shy and furtive bushbuck along the riparian areas are completely nil. We have seen kudu breach the border fence, so we KNOW they are not impressed with the condition of the fence. No matter, creatures are also filtering IN through the same fence.

Steyn introduces us to his camp managers Susan and Evert Potgieter. They are his parents and they are doing a bang up job of taking care of us. We are

What's Left to Hunt?

the only clients in camp! The camp itself is spectacularly situated right on the bank of the Limpopo. I don't mean "by" the river, I mean "on" the river. From the dining area, you can brush crumbs from your lap into the Limpopo. We can sit in the thatched shade of the lapa and soak up the essence of Africa! From the deck there, or even the veranda outside our room, we can watch (and hear) waterbuck, bushbuck, hippo and even crocodiles! Some of these beasts are actually in Botswana. The Limpopo is the geo-political boundary between South Africa and Botswana. The wildlife are not impressed by international boundaries.

After seeing our room, unpacking and getting situated, it's time to meet the rifle I will be using and get sighted-in. I will be shooting Steyn's CZ Safari Classic in .375 H&H. The .375 is the minimum caliber for hunting dangerous game in South Africa. At the range I shoot uncharacteristically well and sense that Steyn is relieved. As a guide myself, I know that the ritual of sighting in is as much about watching my client handle the firearm as it is actually sighting in the weapon. I'm glad I comported myself well and seem to have passed his test. Back at camp we sit around the fire pit and braai, sip cold soda pop and iced tea and nibble on delectable hors d'oeuvres from Susan's well-stocked kitchen. Supper is an eland pot roast and rice. We are in bed by 9 pm; it has been a long day.

The next day Em begins her quest for a Greater kudu. I walk the riparian brush with Steyn and tracker, Patrick, searching for bushbuck. Among other things, we get to see TWO of the "Dangerous Seven" (hippo and crocodile) in the course of our meanderings. Back in for lunch and a nap. Back out around 3 pm. Em accompanies Patrick to the blind at Windmill Pan. They see nothing more interesting than a couple of common duiker. Steyn and I see a few bushbuck, but nothing spectacular. Back at camp we eat dinner and listen to the night sounds of Africa in the river. Steyn gets the "big torch" (flashlight) and shines up a waterbuck and two crocs! This is just so very cool!

Day Two is overcast and 58 degrees. We make a drive for kudu in the morning and check the various trail cams. A lot of different species of wildlife are showing themselves on camera, but the kudu are shy. We head in

What's Left to Hunt?

for a magnificent brunch (eggs, bacon, tomato and corn fritters) and Evert delights in showing me on a map where I will be heading for the buffalo hunt. Looking at the map I can see that the journey down will be a solid 400 km. We have to head back down to Polokwane and then south and east all the way to the outskirts of Kruger Park. Evert explains that we will travel through a famous agricultural area. The crops are primarily citrus and mangos. He says there are lots of "rich people" there. OK by me..... as long as there are lots of buffalo as well.

At noon we load up in the Rover and head out. As we get closer to Polokwane, the road and its Sunday traffic become a complete zoo. To make matters worse, Route 11 is suffering through an extensive construction project. It's after 3 pm when we finally reach the south end. The worst traffic is around the black townships where there are hundreds of people, lots of vehicles in various states of disrepair, donkeys, sheep, goats, cattle and hordes of small children darting onto the shoulders of the road. And frankly, a lot of drivers who are either inebriated or don't know much about driving and don't care to follow the rules of the road.

On the drive down, Steyn and I swap tales about "PH-ing." It turns out that hunters around the world are a lot alike. They often behave alike. We also swap stories of general interest, such as Steyn's friend who misinterpreted the lyrics to the American song "Under the Boardwalk." While The Drifters may look like Metebele warriors to some, for years this fellow thought they were singing "Under the Wart Hog...!" Good times and the Africanization of American rock n' roll! You gotta love it!

We buy fuel for the Rover in Tzaneen and meet our contact "Johan" at the local KFC. "Yes, Virginia, South Africa has KFC!" Johan purports to be the caretaker for the property where we are headed. I note that he is barefoot! Steyn does not seem to have met Johan in person. Perhaps their contacts were all via phone? Steyn quotes his father: "Some people are wise. Some people are otherwise..." It seems to apply here. Johan is personable, but a tad strange.

Through Hoedspruit and out the other side. On toward White River, then Orpen; one of the main entrance gates to Kruger Park is near here.

What's Left to Hunt?

Modern Africa has everything… even Kentucky Fried Chicken.

The entrance to the property is not clearly marked. We learn this as our intrepid guide Johan drives right by it. He discovers his error after a few miles and we turn back and find it the second time through. We are now on "Sandringham," our new home for a few days. We travel a few kilometers on an internal dirt road that has definitely seen better days. Johan misses the next turnoff, this one to the lodge, overshooting it and proceeding an extra two km. I'm wondering what the qualifications are for a caretaker here! Low standards are better than no standards I suppose.

We eventually pull up to the front of what at one time must have been an incredibly magnificent lodge. A handful of what must have been a tremendous staff at one time greet us on the veranda and in the lapa. Ol' Sandringham is seriously run down and neglected. Not a big deal if the buffalo are here, but it is, charitably, an interesting start to our buffalo hunt.

I'm shown to my room. I have my own rondoval, several meters from the main lodge, on the edge of what I would call "jungle." Rosie the maid shows me the interior. It is clean and quite nice, with ensuite toilet and shower.

What's Left to Hunt?

The domestic tabby cat that in the dark, had me convinced he was a leopard.

Not too shabby as long as the lights are low. As she leaves out the front door into the growing dusk, Rosie turns and cautions: "Do not wander outside at night sir. There are lions and leopards about!" Well, boy howdy. You don't have to tell me twice not to wander! As soon as I retrieve my gear, my fat American ass will be staying behind my locked door. Right where I belong!

But first I fire up my headlamp and go back to the bakkie for the remainder of my kit. I make my way back from the parking shed, now in pitch black except for the pitiful beam of my headlamp. This light seemed perfectly robust and bright when I tried it at home in my living room. But here in lion and leopard habitat, it is pretty puny. Back at my rondoval, I climb the front steps to the little veranda. For a moment I stand on the veranda and soak in the sounds, smells and essence of wild Africa. Then I feel something furry moving against my right leg. Holy shit! I'm a goner! I should have listened to Rosie...

It turns out to be only a domestic tabby cat from the lodge! Once my breathing and blood pressure return to normal levels, I make my way back over to the lapa. Steyn, Johan and I sit around the fire and tell a few tales. Mr. Johan is in fact a PH himself. When Steyn leaves for a few minutes, Johan romances me and tries to convince me that I should book with him for hippo, lion, leopard or elephant over in Mozambique. Steyn works in

What's Left to Hunt?

Mozambique himself at times. As colorful as ol' Johan is, I can't get over the fact that it took him two tries to find the gate and that he is still not wearing any shoes. I think I will stick with the known quantity that Steyn provides.

Johan, bless his odd little heart, does however know the history of this property and is willing to share some of it. Sandringham is huge. I never glean exactly how big, but it is in the hundreds of thousands of acres. Only three sides are game-fenced. The fourth side is the border with Kruger Park. It has had low-fence in the past, but the park elephants have shuffled over and through it so many times that it is essentially unfenced now. Game comes and goes from Kruger as it pleases. This of course explains why the buffalo hunting here is so terrific. Currently and normally, there are between 400 and 700 buffalo on the property! At one time, Sandringham was a private game reserve owned by what I understand was an Italian of noble stature and importance. While the buildings have fallen into disrepair and the pool is a nasty mosquito ranch, the wildlife is still here. Sandringham now is part of a conservancy and is for all intents and purposes, part and parcel of the Timbavati sector of Kruger Park. This is clearly going to be a fine adventure. Hell, it already IS a fine adventure. And I haven't even seen my first buffalo yet.

Johan oozes confidence that we will get a buffalo on the morrow. I have already got him pegged as a three-star bullshitter, but secretly I hope he's right! I would not mind if it only took one day to get my Cape buffalo bull.

Patrick removes some hot coals from the fire pit and readies the grill for lamb chops and sausages. Rosie fetches pap (cooked maize meal) with a tomato sauce, as well as potato salad, from the kitchen. We sit down to a great meal while being serenaded by hyenas in the distance. After supper I sign off and promise to be back in the lapa for breakfast at 6 am. Sure will! If I can sleep at all!

The next day dawns overcast and 58 degrees. Sometime after midnight, all the power went out in my rondoval. I thought that maybe they just turned off the camp generator, but alas, this is not the case. It is still "lights out" when I get up at 5:30 am to begin dressing by the Braille

What's Left to Hunt?

In perhaps my all-time favorite analogy and quote, writer Robert Ruark described the glower of a Cape buffalo bull: "He looks at you as if you owe him money."

method. Having only my trusty headlamp to assist me in getting dressed and organized, I'm glad I laid out my gear last night.

I slept only fitfully. I am JUICED to hunt Cape buffalo! Then of course, at 4:30 am lions began roaring somewhere close by. Being kept up at night by a snoring spouse is amateur stuff. Let me assure you, you do not remain asleep when lions are roaring in the neighborhood. This is positively awesome!

At 5:55 am, I don my headlamp and cautiously (VERY cautiously to avoid the lions and leopards don't you know...) make my way over to the lapa for muffins and coffee. As the sun comes up, I witness some "monkey larceny." There are vervet monkeys all over the place. Apparently no one here owns a pellet gun or a .22 LR! Anything not nailed down in the dining room is scooped up and hauled off by one or more monkeys. I selfishly guard my breakfast fare. I'm able to down a muffin, some rusk and a coffee without any pilfering by the busy monkeys.

What's Left to Hunt?

Vervet monkey sneaking into the lapa intent on stealing whatever is not nailed down.

Steyn proclaims the lions I heard were females. He may be blowing smoke, but he sounds convincing to me. Might even be true. Steyn says the males pretty much stay in the park and do not come out onto the surrounding private reserves. It's a survival thing.

I sit and watch a magnificent African sunrise to the east, which is to say over toward Kruger Park. I'm treated to the sight of ducks, herons, geese, and fish eagles. I observe one significant difference between this buffalo camp and those of yesteryear that I have read so much about. Instead of briefing their client on the nuances of buffalo hunting, both PH's and our intrepid tracker are this morning checking their email and reading their texts on their smart phones! This is African Safari 2016, baby. The New Millennium. The Brave New World. Obviously Theodore Roosevelt, Robert Ruark and Ernest Hemingway didn't have to put up with that shit! But I am in heaven anyway. Roaring lions, marauding monkeys and a young bushbuck frolicking on the lawn. This is positively incredible! Let the staff have their smart phones. No matter.

What's Left to Hunt?

An ancient touring car arrives with Sandringham tracker, Ben, at the wheel. Ben is dressed like a doorman or a chef, but it soon becomes clear that he has been on the property forever and is the real deal. Both PH's, Patrick and I all load up on the back of the bakkie. We launch and are soon driving along a muddy braided river bottom. Steyn proclaims for my benefit: "This is where the dagga boys get their dagga....!" What a great line. "Dagga boys" are a reference to bachelor buffalo bulls, who often roll in dagga (mud) and coat their hides as protection from biting insects and external parasites.

By 7:30 am we are looking at our first two buffalo. Steyn puts up his binoculars and proclaims "one of these will be just fine." But before I can get ready to shoot, the bulls wheel and run. We hop on their track and move along behind them. While we do spot them again, the wind is swirling and the bulls do not stick around. Ben tracks them until they split. He makes the decision that they are "moving country" and with so many buffalo available, it is of little use to follow these particular animals. Back to the bakkie to round up more tracks or animals.

The next couple of hours are spent looking at a fair amount of countryside. We see a female hippo, some eight different giraffe of various age classes, and a small group of cows and immature bulls. Later we find another five giraffe and finally, a lone Cape buffalo bull.

Ben whispers that he knows this bull and that he is a good one. The two PH's evaluate the bull quickly, judge it to be good, and advise me to shoot. At the magic words "Take him....," I squeeze the trigger. Amateur mistake; the safety is still on! Crap. My reputation is shot and the buffalo is NOT shot....! I flip the safety off and squeeze off what I would say is the perfect shot. Right out of a Daktari Robertson's "The Perfect Shot" instructional picture book. Shoulder/high lung at the broadside bull at about 65 yards. Hit solidly with the .375 H&H, the bull humps up and lunges into thick scrub. Steyn had enough faith in me that he was holding binoculars instead of his .458. He says he clearly saw where the mud flew from the bull's shoulder and that the shot is, in his considered professional opinion, "good." I have of course cycled in another .375 round, but have no chance for a follow-up shot as the bull disappears in the brush. Steyn and Patrick are shaking my hand, so I labor under the assumption that I have done "OK," and perhaps even "Good." I guess we will see.

What's Left to Hunt?

This is why older Cape Buffalo are called "dagga boys." They cover themselves in mud.

We wait about 15 minutes, then start into the brush after the bull. We go less than 40 yards when Patrick spots our quarry. The bull is hit hard, but still on his feet. I come up to Patrick and get two more shots into the animal. The first round looks pretty good. The second round is a bit far back as the bull crashes away in the thick stuff. But this time he goes down. Steyn also puts a .458 slug into his right shoulder. We advance another five yards. I kneel beside Steyn and put two more in him. The bull is down, but still "high-headed" and not finished.

Johan circles out to the left. Suddenly the bull lurches to his feet, glowers at Johan and takes a couple of steps toward him. Johan brings up his own .458 and puts one onto the bridge of the bull's nose at 10 yards. This puts him down for good. In short order we are done. "No hunters or trackers were harmed in the making of this story…" In my opinion, that is the measure of a successful Cape buffalo hunt.

What's Left to Hunt?

Pete, Steyn and Patrick with the Dagga Boy. A lifelong goal achieved.

Now the real handshaking begins. About the time the hand shaking is done and I'm getting out two cameras, the adrenaline shakes kick in. For a few minutes I can't even operate my own camera! But better I should get the adrenaline shakes now, rather than during hot pursuit. Other than trying to fire the .375 while the safety was on, I feel I have done pretty well during the important part of this drama.

Ben brings up the bakkie, runs out the winch and begins preparations for loading. We take a ton of photos. The bull is quite old, hard-bossed and is a fine trophy. Ben has seen him on this part of the property for quite some time. Several hunters have tried to get him, but all were outsmarted. Ol' Pete has been lucky. I'm very appreciative of my luck.

Back at the lodge we partake of another fine brunch. I have a calm sense of accomplishment. This hunt has been on my bucket list for ages. I've long wanted to achieve this pinnacle. I have done it. I have a tremendous trophy "in the salt." No one got hurt. Tomorrow we will travel back to the Limpopo Province. Life is good.

Chapter 4

"Plains Game" – At 7,000 Feet

The Common eland is an iconic African antelope. It is also the largest antelope in the world. The Cape eland is a subspecies of the "common." A large Cape eland bull can weigh more than a ton and stand six feet at the shoulder. Eland are not considered "dangerous" game, nor is hunting them generally considered particularly arduous. They commonly inhabit arid, primarily level terrain in the mopane scrub and thus are classified as "plains game." Taking a big eland is often accomplished by walking and tracking great distances. But seldom is mountaineering a part of the skill-set required to take a big "blue bull" as the older bulls are often called. I have a penchant for never doing things the easy way. My Cape eland bull may have technically been mere "plains game," but he was living on high!

My PH, John "Madala" Barnes, of St. Francis Bay, RSA, loaded me into his vintage "Landy" (Land Rover) one morning in the Eastern Cape in the pitch dark. We stopped and picked up a substitute tracker named Rasta (our regular tracker, Leander, was bound for a clinic in Grahamstown for some dental work today) and headed out to a property near the town of Cradock. Cradock lies along the upper reaches of the Great Fish River (Groot-visrivier.) It is the administrative seat of the Inxuba Yethemba municipality in the Chris Hani District. Founded in 1813 to cater to the migratory farmers who settled this rugged and hauntingly beautiful area, Cradock supports many of the thriving cattle businesses to this day. Game parks and eco-tourism have become popular lately, but commercial hunting also remains viable and important to property owners.

We make it from Waterfall Farm out to the pavement in short order. On through Bedford, Cook House, East Somerset and Pearston. The names harken back to the English settlers of this part of the Eastern Cape. Finding the next turn is problematic. We prove this by missing it and end up backtracking nearly 10 km. We ease up through the narrow valley of the Great Fish River, working our way back into the mountains and quite higher in elevation. We pull into the property we will be hunting around 8:30 am.

"Plains Game" at 7,000 Feet

The property is called Asante Sana, which I find interesting. "Asante sana" is Swahili for "thank you so very much." I've not heard much Swahili spoken anywhere in South Africa. English to be sure. Afrikaans of course. And primarily Tswana and Xhosa. The only Swahili I detect is on the gate sign.

Just inside the gate we encounter a delightful couple, Richard and Kitty. They are walking a pair of dogs and .... a warthog! Turns out they are the property managers and so we stop and make our manners. Kitty jumps in with us. Richard walks on with the dogs and the warthog. We motor up to the office to sign the liability waivers for use of the property. Kitty explains that this formality is necessary for "if we shoot each other!" I rather imagine it is in case we get the poop stomped out of us by one of the several elephants or rhino that are on the property. She reminds us to stop back if we need anything, but warns us not to get out of the vehicle. "Just park here and 'hoot' (honk the horn...)" I'll come right out." she cautions. "The pig can be aggressive!" This is actually the very first office complex that I have seen, anywhere, that is guarded by an attack warthog! Richard, dogs and "pig" soon catch up.

John and Richard hit it off immediately. Both have discerning taste in older Land Rovers. Richard asks as to whether John has "as yet had the engine replaced?" No!

We move over to the shop and pick up our local tracker, Able, a wizened San Bushman. I guess him to be in his mid-40's, but it is very hard for my Caucasian mind to accurately ascertain his actual age. The conservation laws in South Africa dictate that when hunting a property, you must have an employee of that property with you at all times. When you are in completely new territory, having a local guy is more than just handy; it's essential. In this case, Able has been on this property for ages and knows it like the back of his hand; he will prove indispensable before the day is over. Did I mention that he speaks no English at all? Minor item; no problem. John is of course fluent in Afrikaans; he and Able have that language in common. Rasta speaks Xhosa and some English. I speak English only; I know some Afrikaans words and phrases and just a word or two of Xhosa from previous trips. Communication today will be an exercise in patience and perhaps excitement at times!

"Plains Game" at 7,000 Feet

Band of eland bulls browsing their way across the mountainside.

With Able pointing the way, we drive out of the barnyard and up toward the formidable Sneeuberg Mountains. Apparently Able has told John that the "plains game" (eland) that we seek prefers the "somewhat higher elevations" of the Asante Sana property. Before leaving the tilled fields of the riparian area on the valley floor, we are lucky enough to spot herds of impala and red lechwe. The lechwe are a first for me. It is a very beautiful antelope. Up the mountain we proceed. At each track intersection, Able points the way. This particular track seems to be more and more in need of maintenance as we climb. For a while I assume that we are merely heading for a high observation point and will be glassing down at the lower elevations for eland. After all, eland are "plains game!" Certainly plains game do not live at these elevations, do they? As it turns out they do indeed. We observe black wildebeest, blue wildebeest, red hartebeest, mountain zebra and even a rhino cow and calf at some unbelievable elevations! Far up in the rocks. Essentially above timberline.

"Plains Game" at 7,000 Feet

We park the bakkie at a dilapidated old developed-water point and begin climbing ever higher on foot. After 30 minutes we are nearing timberline. By this time I suspect Able is blowing smoke insofar as eland are concerned and probably getting a large kick out of wearing out the visitors. As the temperature rises, the winds become almost completely convective. They are blowing our scent directly up drainage. If any eland are above us, they surely have our scent by now. But, bless his heart, Able is pointing out more and more eland sign! By 11 am we are far up the mountain. Able points out four cow eland on the mountainside across the drainage (and a bit higher in elevation!) I would not have believed how far up the mountainside they are lounging. But there they are... plains game in mountain sheep country!

Moments later, from the same glassing point, we observe what we have been seeking... a bachelor group of eland bulls. They are feeding in the brush growing around the rocks. There could be five of them, but they too are on the opposite mountainside, all of 700-800 yards away. There is sparse to no cover between us though. For now we must be content to just observe them. There is no good way to sneak closer to their current position.

John queries Able as to what his plan might be. A conversation in Afrikaans ensues. Eventually John translates the essentials of said plan from Afrikaans into English for me. According to Able, he KNEW the eland bulls would be there! Able has told John that he sent two ranch hands, on horseback no less, to come up the back of the mountain on which the eland are now browsing. The plan, such as it is, is that the African cowboys will appear a bit later this morning, in the saddle above the eland bulls. "Perhaps" the bulls will move our way! This sounds a bit shaky to me, but my policy is to never question my guide. From a practical standpoint, I subscribe to a policy of not attempting to "guide the guide." The corollary is a stiff dose of "When in Rome, do as the Romans do." This is Able's home country and it is his show. I have nagging doubts about whether eland are any more conducive to being steered or herded than say, caribou are. But I have every intention of doing exactly as Able politely "suggests." In a valiant effort to keep his client's spirits up, John begins referring to the still unseen riders, as the "Horsemen of the Apocalypse!" I wryly observe that this is definitely some badass country to move about on horseback. Frankly I will be shocked if riders appear in

"Plains Game" at 7,000 Feet

that pass. And even more shocked if they manage to push any eland in our direction. This is mainly because there are a lot of different directions in which they could move, without coming anywhere near us.

Thus we sit in the bright sun, tolerating the heat only because of the brief respite offered by a brisk up-canyon wind. We see some other wildlife, including a huge waterbuck. We are treated to an occasional alarm bark from a baboon who is well-convinced that we do not belong in his valley. The longer we sit, the less I am convinced there are any horsemen. Apocalyptic or otherwise!

But what there are now, are a total of NINE eland bulls in sight. A couple appear to be very large bulls. Deep "blue" color, with "poofy" brown forehead hair. They browse contently through the scrub and rocks. When they have all moved across the mountainside and out of sight, we are able to make our move across some of the open country between us. We have been watching them for nearly two hours. Now we can move across the virtually cover-free valley bottom. Off we go, with Able in the lead. We make great progress in closing the distance. Soon we find ourselves 360 yards from the eland bulls. The animals are back in the open and we are pinned down once again. John and I discuss my attempting to make the shot from here. There are some issues to be considered. First, eland are huge, with a kill zone the size of a big wash tub. I am armed with John's M77 Ruger in .300 WinMag; this is a shade on the light side for a 2,000 pound animal, but an accurately placed bullet will do the job. Second, I know my own limitations. I am not terribly comfortable shooting at this distance, but I can. John has seen me shoot; he guided me on my 2013 safari in the Eastern Cape, using this same Ruger in .300 WinMag. He is confident in my ability to pull the shot off, but says he will not pressure me if I am uncomfortable with the shot opportunity presented.

I settle in, standing upright, with the .300 resting comfortably in the crotch of a tree. It is a solid shooting position. I lock down behind the rifle and dry fire it a time or two. I am now confident I can make the shot and kill the animal at this range, under these conditions. Ironically, when I am finally ready to make the actual shot, the biggest bull is no longer showing

"Plains Game" at 7,000 Feet

As promised by our tracker, Able, the local horsemen really did show up in the pass above the eland.

in the open. Before the big one shows again, they ALL move off around the mountain! We take advantage of the opportunity to move closer. We are lucky to be able to get across another large opening without spooking any of the bulls.

John jokes that we should "wait for the horses...!" I look him in the eye: "There are no horses, are there, Madala?" He just laughs. I'm not sure he is convinced either!

Now we have made it to a little kopje (hill) below where the eland are feeding. We are on their side of the drainage, thus they are above us at a fairly steep angle. My rangefinder indicates we are now at a more reasonable range, 242 yards. There is no way to utilize a prone, or even a sitting or kneeling shooting position. Although standing, I do have the shooting sticks firmly planted. I'm very comfortable with this setup. I can make the shot.

"Plains Game" at 7,000 Feet

John "Madala" Barnes and Able the San Bushman tracker pause on our way off the mountain.

When our chosen quarry steps clear of the scrub and is not lined up with any of the other animals, I am ready. I carefully squeeze off a round. The .300 barks; the bull lurches and flops over into thicker brush. Able signifies through hand signals his confidence that I hit him well. I chamber another cartridge and re-adjust the sticks. When he shows again and is still standing, I send the second round. It is all over. We start the steep climb. Rasta gets to the bull first and yells back down to assure us that the bull is stone dead. And what a bull it is. Judging from John and Able's reactions, this is a special trophy. Having never hunted, much less shot, an eland, I am not a good judge of just how good it might be. Everyone else's level of elation is catching though. I am pleased with myself. A great trophy in an incredible spot, after a well-planned stalk, in country that is more like Dall Sheep habitat than anywhere I envisioned finding "plains" game.

Reality begins to set in. The bull is lying in a terrible spot in terms of ease of recovery. With all four of us straining, we are barely able to roll the bull

"Plains Game" at 7,000 Feet

The iconic white rhino, a terrific conservation success story in sub-Saharan Africa. Another species saved by hunters and hunting.

up onto his brisket for a properly respectful photo. Able dives into the butchering process with a vengeance. He disjoints the four feet and tail and stashes them for himself! John and Able and I begin to cape the bull.

I am well convinced that we are in the middle of nowhere. But within minutes we hear a faint shout. Who could it be? Here come the aforementioned horsemen! As it turns out, there really were horses! I should never have doubted Able. I am intrigued by the gear on these Sub-Saharan African cowboys. One rider has leather leg gaiters that stretch from ankle to knee. The second fellow has similar gear, but his gaiters are of heavy felt. This protective gear is obviously designed to protect the riders' legs from brush, cactus and thorns. It dawns on me that everyone in this hunting party is wearing long pants.... EXCEPT the American client. I am clad in shorts and my legs are a bit the worse for wear from pushing through thorn brush all morning. There may be a lesson lurking in this observation!

"Plains Game" at 7,000 Feet

The trophy "blue bull" where he fell on the mountain.

Able assures us help is on the way. I have learned now not to doubt this intrepid little San Bushman, so John, Able, Rasta, and I head off down the mountain. It is quite steep and my knees are talking to me long before we reach the Land Rover. We haven't gotten far before I see the "help" that Able promised. Coming up the rough track is an ancient farm tractor, popping and belching diesel fumes. On the tractor and the flatbed trailer hooked behind it are about 15 indigenous personnel from this farm. Two fellows are riding on the blade on the front of the tractor. Every time the tractor goes across a dip or ditch, these two guys are flung into the air. Sometimes they come back down in the bucket again. Occasionally they don't. The "misses" are jeered loudly by their compadres riding on the trailer. I must admit, when I first observe this crew coming up the mountain, I am well convinced that I am witnessing a prison break of some sort. They are a rough-looking bunch. Various bits of ragged clothing, assorted hair styles (from "fros" to shaven heads), tattoos, missing digits and missing teeth! This is one raggedy crew. From lower on the opposite mountainside I observe that they are able to get the tractor and trailer to within a half mile of the kill site by maneuvering it across some tricky eroded ground. From there, says Able, they will walk

"Plains Game" at 7,000 Feet

The staff from Asante Sana farm that constitute the "recovery crew" showed up to bring my eland off the mountain in fine form.

up, subdivide the eland into chunks of 30-50 pounds and hand-carry them down to the trailer. I could use these guys the next time I kill a moose in a bad spot in Alaska, but the immigration issues could be tricky!

We leave the kill site at about 2:30 pm and arrive back down at the bakkie in a little more than an hour. We had left Kevin's farm before first light this morning to drive to Cradock. John and I are pretty much "running on empty." We pounce on the "chiller box" and devour some of Natalie and Trust's fine sandwiches, washed down with ice-cold Cokes. Able gratefully joins us. Then we rest in the shade awaiting the recovery crew. They show up amazingly soon. This is clearly not the first eland that they have butchered. Fifteen sets of experienced hands have done a bang-up job of cutting up and packing out my eland. We make our way down to the homestead, check in with Richard and Kitty and hit the road back to Waterfall Farm. A great day that has resulted in bringing a fine specimen of the world's largest antelope species to bag. Plains game taken at just shy of 7,000 feet elevation with cowboy support, no less!

Chapter 5
How to Draw an Alaskan Bison Permit

In 2006, I took stock of my luck in drawing permits to hunt, in Alaska and other states. It turns out I had unsuccessfully applied to be drawn for a permit to hunt an Alaskan bison for nearly 40 years! Yes, Alaska has free-range Plains bison. The basis for the original herd were brought from Yellowstone back in 1926 and released near Big Delta in Interior Alaska. Since then, ADFG has stocked other suitable habitats. We now have three areas with small annual huntable surpluses. Delta of course, but also the upper Copper River/Chitna country, and the Farewell herd which ranges from the Middle Fork of the Kuskokwim, north to the Farewell burn and up the South Fork to the Post and Hartman River drainages. That's the good news.

The bad news is that the odds of drawing a bison permit were and are, abysmally low. For example, in 2020, there were 172 permits offered. But nearly 50,000 applications were filed for those 172 permits. That is a less than 1% chance of drawing any given bison permit. Alaska has no "preference point" system. Your odds are the same the first year you apply as when you apply in Year 40. I estimated that I had spent close to $1,500 applying for permits on behalf of myself, my wife, my son (and my dog; apparently he was not lucky either.) The only luck I had experienced was in not getting caught putting in applications in the dog's name. A lot of people die without drawing an Alaska bison permit. It was starting to look like I would be in this category. I probably could have bought a pretty good amount of freezer beef for less than I had "invested" in bison permit applications. Clearly I was not going to draw a permit any time soon.

Oh, I had HUNTED bison. But with other people who had been lucky enough to draw permits. In fact, when I initially became an Alaska Registered Guide, the very first hunt contract I ever executed was to guide the 13 year old daughter of a Fairbanks physician. With my assistance, young Beth took a huge, B&C class bull. Years later I guided another Fairbanks doc to his own bison trophy. And during the interim, I accompanied other successful

How to Draw an Alaskan Bison Permit

Herds of bison still roam the range. On the Crow Reservation, the Crows refer to them as buffalo!

permittees. I had hunted for bison. I had spotted bison. I had stalked up on bison. I had photographed bison. I had gutted and skinned bison. I had butchered bison. I had eaten bison meat. And lord knows, I had packed loads of bison meat on my pack board. But I never killed a bison. I could not draw a bison permit myself. No way. No how...

In the summer of 2006 I became aware of a bison hunt being raffled off to raise money for the support of the Hardin (Montana) Elementary School Swim Team, on the Crow Reservation at the foot of the Bighorn Mountains of southcentral Montana. "Go Otters....!" A trapper buddy in Hardin had a daughter on the Swim Team. Alan tricked me into sending along a few dollars and purchasing tickets for the bison hunt raffle to help Nicole's swim team. Go Otters! Go Pete!

Considering my abysmal luck in drawing bison permits, I soon forgot about the raffle tickets I had purchased in summer, but in mid-September I got a call from Montana. I had won the "free" buffalo hunt in Montana! My luck

How to Draw an Alaskan Bison Permit

was changing, or so I was starting to believe. I had finally bucked impossible odds. What possibly could go wrong now? I would find out soon enough!

In late October I flew to Billings, rented a car and drove down to Hardin. I met up with U.S.D.A. Forest Service friend Frank. Frank drove over from Custer, South Dakota to video the hunt. But alas, there was no hunt to film! Thanks to some terrible weather, the hunt was not to be. We spent the week weathered-in at our motel in Hardin, waiting for a raging Rocky Mountain blizzard to subside. It didn't. Frank drove back to South Dakota and went back to work. I flew back to Fairbanks. The Crow Tribe however, graciously invited me back to try again the following year. They cautioned me to try earlier in the season. I would try again in September 2007.

This time, several weeks earlier in the fall, the weather was much milder. Frank had been smart enough not to sign on for videographer duty this time! I had visited a couple of college buddies in Billings and then driven down to Hardin and Crow Agency. This time, the Weather Goddess was smiling down on me...

September 26 is a bright, sunny fall day. I get my .375 H&H sighted in at the local gun club range, then head to Crow Agency and meet with Alvin, the head of the tribe's Buffalo Project, do final permit paperwork and make final arrangements for the hunt. The Crows call them buffalo, not bison. When in Rome...

After the visit to tribal headquarters, I spend the rest of the day wandering the Custer Battlefield National Monument. After supper at the ubiquitous Purple Cow (Hardin's only hotspot....!) it's back to the "Second Best" Western Motel where I meet with trapper buddy Alan, himself a resident of the "Greater Hardin Metropolitan area!" Alan had sold me the winning ticket and was still pretty proud of himself. In addition, Alan's church has a permit from the tribe for a meat buffalo; Alan is the designated shooter. Tomorrow will be a hunt for two adult buffalo. A dry cow for the church and a nice bull for me.

How to Draw an Alaskan Bison Permit

This bull is the perfect size. Big enough to have nice horns. Young enough to be great table fare!

It is a full month earlier to the day than when I was here last year. But it is not snowing madly. This morning it is 46 degrees and crystal clear. I gobble down some breakfast and head over to Alan's house. In his 4WD truck we head down to Lodge Grass where we meet our intrepid band of Crow buffalo hunters at the Cenex Station, then on south toward the Big Horn Mountains and the highest elevations of the Crow Reservation. In the lead pickup are the Crows. David, Loris, Eugene and Henry. All four guys are current or former fellow wildland firefighters. We have a ton in common. We come to find out that we have even been on some of the same fires over the past 20 years.

We leave the pavement and start making our way up the mountain under the light of a full harvest moon. There is some snow, but nowhere near what we battled last year. As daylight breaks, we pass the highest point we got to last year. Onward we go.

How to Draw an Alaskan Bison Permit

A bit later, a nice six point bull elk trots across the two track in front of us and stops to peer at our little convoy. The Crows pour out of the pickup and lay down a withering field of fire with their beater "truck rifle," an ancient 270 Winchester. The elk decides he has made a terrible mistake by showing himself to the denizens of that pickup. He hauls ass for the Wyoming border to the south. Loris is able to get off two magazines worth of .270 Win rounds, but to no avail. Nary is a hair dislodged on the bull. He is untouched, but smarter. He continues his "run for the border." We keep heading to where our buffalo should be waiting.

We make Windy Pass, which at 9,100 feet is the highest point on our route. We are above timberline and into an area of lush grasses so important to the maintenance of the tribe's buffalo herd. It is fantastic habitat. By 9:30 am we are perched high above the "Buffalo Pasture" and watching a lot of buffalo! This is like a scene from a Charles M. Russell print. These are happy buffalo, grazing contentedly high above the steep canyon walls that fall off down to the Big Horn River and the Yellowtail Dam far below.

In the closest groups there are dry cows that would fill the bill for Alan's permit, but no bulls for me. We leave the pickups and hike along the rolling plateau. There is minimal cover except for terrain changes. If we stay in plain sight, any time we get to within rifle range, the animals leave for safer ground. Their comfort zone seems to be about 300 yards, which I find a little far for me to shoot at a large, tough animal.

And so we ease along, looking over small groups of buffalo. Finally we crest a hill and find ourselves within 80 yards of a more diverse group of animals. There are a couple of suitable bulls in this bunch. Loris points them out to me and I select one. The bull looks to be seven or eight years old. Big enough to make a nice trophy and young enough that he will provide great table fare. Perfect choice!

But it is not as simple as just shooting this buffalo. I must wait until he is clear of all other animals. If there is a cow or calf in front of or behind him, I cannot chance a shot. A pass-through is not likely on such a large animal, even with the .375 H&H, but I cannot take a chance of wounding another buffalo. It seems like an age, but finally the bull I have selected is in the open.

How to Draw an Alaskan Bison Permit

My seven or eight year old bull. Taken cleanly with the .375 H&H. Now the fun begins!

I have the rifle on the sticks and most of the slack out of the trigger. I hold for just behind the shoulder and squeeze off a round.

At the report and hit, the bull turns and walks less than 10 yards before tipping over. From the time he was hit, he remains alive for only another 10 or 12 seconds. Later examination shows that the 300 grain Nosler neatly clipped the dorsal aorta from the top of the heart and punctured lower portions of both lungs. The bullet did indeed pass all the way through the animal. In fact, it never even hit a rib, either entering or exiting. Minimal meat damage and a very rapid kill.

The rest of the herd begin to file off, but they really aren't terribly disturbed. A dry cow is last in the procession. Alan kneels and gets steady with his rifle. When the cow momentarily stops, Alan puts a .338 WinMag round through her neck, just behind the ear. She drops like a box of rocks. We are "done." The easy part is over. We go back for the trucks and carefully work them down to the animals we have taken.

How to Draw an Alaskan Bison Permit

Any time you have this many experienced hunters helping out, field dressing and quartering goes pretty smoothly.

Loris walks over and shakes my hand. He is an experienced guide and has seen all manner of shooting prowess (and lack of it). He is glad that we shot well and killed our animals quickly and with respect. He is also remembering that he shot up a half box of shells this morning at a bull elk he did not hit. He mentions this, and then quips: "With the way I was shooting this morning at that bull elk, it is a good thing we are not playing cowboys and Indians!" Cracks us all up!

We separate into two teams and begin the butchering work. The Crows clearly have done this a time or two. My years of field dressing and cutting up moose in the wilds of Alaska are serving me in good stead too. This won't take long.

At one point the Crows ask me if they may retain the "soft parts" of my bull. I tell them "of course" since I am already looking at a pretty hefty air freight bill to get my meat back to Alaska. The contents of said "soft parts" are stripped out and bagged up for the trip down the mountain. We lay out tarps in the back of the two pickups and place the meat on them. As with any game, keeping the meat clean and dry and allowing it to cool rapidly, is the key to having great table fare later on.

How to Draw an Alaskan Bison Permit

We start down the mountain, stopping at an old home-site near Commissary Creek. Here a pipe has been driven into the mountainside; fresh spring water pours out. Bloody things are washed. Intestines are rinsed out by clamping them over the end of the pipe and allowing water to flow through. I have been interested in learning some Crow names for things we have been seeing. As talk turns to what the use of the "soft parts" will be, I politely inquire as to how they plan to cook the intestines that have been salvaged. I also ask what the Crows call this campfire treat. Loris replies that they call this "menudo." Instead of just accepting this, I press him further: "But that is Spanish, not Crow. Why do you call it 'menudo'?" With a completely straight face, Loris replies "Because if we called it 'inside of the asshole,' no one would eat it!" Understandable indeed! Once we recover from that overdose of Crow humor, Loris explains further that they normally chop it into smaller pieces and just grill it directly over an open fire. Sounds good. And his explanation is cogent!

By dark we are off the mountain, tired but happy. Alan and I drop the meat of both buffalo at the processor in Hardin and proceed to McDonald's to chow down. I have no trouble sleeping this night.

After a couple of days I pick up the meat. I high-grade out the best cuts of prime buffalo meat and freeze them to be air freighted back to Fairbanks. I also donate the remainder of the meat, already cut and wrapped, back to the tribe for distribution to needy elders.

But remember, this chapter is about how to draw an Alaskan buffalo (bison) permit. Just four months after I kill my Montana bull, the Alaska Department of Fish and Game draws my name for a coveted Delta Bull Bison Permit! After 40 years of unsuccessful applying, as soon as I kill a Montana bull, my luck kicks in and I beat the odds in the Alaska permit draw!

See? Nothing to it! Anyone can do it. And yes, the following winter I hunted Delta, passing up 11 big bulls, and eventually filling my permit with a 20 month old bull bison. If anything, it was even better table fare than my Montana bull! It just didn't have as many Frequent Flyer Miles between the field and the dinner table.

Chapter 6

Fotch Creek Bull

Like most experienced hunters, I realize that once I find a good productive place to hunt, I must be fairly careful not to blab about where that place is located. Once others discover where you are finding success, invariably they go to your spot instead of finding their own. Over the years I have gotten in the habit of being less than precise when asked, "Where did you shoot that moose?" In fact, sometimes when that question is posed, the answer is: "Right behind the shoulder!"

Other times I off-handedly reply: "I got him over in Fotch Creek." If there is one characteristic shared by most Alaska hunters, it is that they pride themselves on knowing every creek, every lake, and every mountain around. So when you mention an imaginary waterway, such as Fotch Creek, they normally accept your answer. They would chance losing their hunting "man card" were they to admit that they do not know where Fotch Creek is. But there is no Fotch Creek. I invented the name!

When I'm hunting for a "meat" moose, I don't deliberately hunt for large antlered bulls. My goal is to find a legal bull, in a place where it won't be difficult to recover it. Over the years I've packed tons (yes, literally tons, which explains the hernia operations...) of moose meat. These days I am unlikely to shoot a bull that is on the downhill side of a logging road or in a place where I cannot get to it with a 4-wheeler, a boat, a horse or whatever other form of transportation I might have at my disposal for moving moose meat.

When guiding, one is more likely to allow moose to die in difficult spots. After all, since one's client wrote a big check in the sincere hope of shooting a big trophy bull, it is incumbent on the guide to let him shoot one. Even if it is found in a less than ideal spot from which to retrieve meat. For a number of years, I worked as part of a group of intrepid moose guides who figured out that using tracked vehicles to get further off the road system than most hunters were willing to go was a pretty slick way to salvage moose meat.

Fotch Creek Bull

Moose tracks in the silt and mud along an Interior Alaska river. Sure evidence of a nearby bull.

I sort of fell into the job after having my first hernia operation (ironically, from packing moose meat on my back). I had been warned by my doctor that I should avoid back-packing moose quarters unless I was interested in experiencing that same level of pain again. I am not a big fan of pain, so I took his advice and threw in with the group with the Bombardier muskeg tractors. These machines go some amazing places and usually don't get stuck. When you DO get one stuck, getting it extracted is a major operation! But I digress...

On this hunt, by late September of 1987 we had finished up with our hunting clients. We languished for a few sunny fall days in the splendid solitude of the Alphabet Hills in GMU 13. We cut and split firewood for the camps, performed some maintenance on cabins and tent platforms, and picked a variety of berries. We packed the gear and some meat into an old military "deuce and a half" trailer with aircraft tires and headed toward the road and civilization. Termination Dust (light snow) covered the hills as we headed out to the Denali Highway. Late morning of the first day we stopped to "pull tracks" on the Bombardier. This process consists of manually yanking

Fotch Creek Bull

Bull moose in late September move a lot in search of cows. It's a great time to hunt for them.

the roots, brush, snow and ice out of the steel tracks of the machine. There were plenty of hands on deck to accomplish the task. My friend Dave and I walked on ahead. We all agreed that we could use another moose for meat for the crew, and the willow patch in the next drainage was a good place to see a moose. We agreed that if a legal bull showed itself, Dave would have the option of the first shot.

As Dave and I meandered through the willow patch, I heard a grunt. A large bull stood up, barely 40 yards off the trail. At that time, a legal bull in GMU 13 had to be 36 inches or better in spread. This bull was clearly WAY bigger than that. "Shoot," I encouraged Dave. But instead of a shot, I heard Dave say "Is he legal?" "Yes, yes; shoot...!" I urged. But Dave was busy. First he was chambering a round in his rifle. Then he was removing the lens covers from his scope. Then, apparently, he was trying to find a moose in his scope. Coincidentally the bull moose was busy too. He was busy fixing to leave the immediate area and he was doing it quickly. I made a fairly quick decision. If we were going to get this moose, I was going to need to shoot it.

Fotch Creek Bull

The bull was headed almost directly away from us, but he was working his way up a steep hill. I held for the back of the base of his neck and hit him between the shoulder blades with my 7mmRemMag. He went down like a box of rocks.

I started into the brush after him. Dave unloaded his rifle and put it on safe and followed me as we walked up on the bull. Other than the fact that it had a lot of meat on it, this animal hardly qualified as a "meat" moose! In fact, to this day, he remains the largest moose I have ever taken, both in body size and in antler size. He was a dandy.

The other guys of course had heard the shot and quickly came on with the Bombardier. With the whole crew working, we make short work of the butchering chores. Soon we were back on the trail out to the road.

A couple of days later I headed back to Fairbanks, arriving home around mid-afternoon. I hung the meat in the garage and stood the antlers up on the end on the porch. Just prior to supper time, my wife Jan came home from work. (Note: Jan was my first wife. She passed away in 2002.) I would have thought that she might say "Nice to have you home from the hills. Wow, that is a lot of nice moose meat you have there." Instead she queried me "What are those antlers doing on the porch?" I told her that it was the biggest moose I had ever shot and that it was my intention to mount the antlers and hang them in the living room. "Nope. Those are not coming in the house," she laid down the law. I moved the antlers into the garage and began trying to figure out a way to circumvent the spousal edict.

My chance came the following March. We were in town at the J.C. Penney store. Jan spotted a living room set that she coveted. I told her, "Go ahead, buy it." "Can't," she answered "I don't have enough money in the household account." "But," she cheerfully suggested "You have that much money in your business account, don't you?"

"Sure." I said. "I'll make you a deal!" I'll buy you that living room set, if you will let me have my European mount moose antlers in the living room!" She agreed; she really wanted that living room set! I wrote a check for the

Fotch Creek Bull

Posing with the Fotch Creek Bull. This bull is likely the largest body size of any moose I've killed in many years of moose hunting.

furniture. She wanted the sofa in a different color fabric than the one on the showroom floor, so it would be a month or so for one in the desired color to be delivered. That gave me time to make a really nice plaque and fancy up the antlers of the Fotch Creek bull. In a couple of weeks the antlers were ready to hang up, but I wisely waited.

Finally the Penny's warehouse called and told me the delivery would be the following afternoon. I took the afternoon off from work and drove home. While I waited for the delivery truck, I hung my big ol' moose antlers in the gable end of the living room. They looked GREAT there too! Soon after that task was accomplished, the warehouse truck pulled up and the delivery guys moved the spiffy new living room set into the living room. I must admit, it looked GREAT in there too. I sat back in the new recliner and awaited the arrival of the Queen of the House.

Fotch Creek Bull

The Bombardier Muskeg Tractor is an Alaskan moose hunter's pack horse. Saves wear and tear on your body and packboard!

A bit after 5 pm, Jan walked in. I smugly awaited the expected words of praise about the furniture. But instead of complimenting me on how great the new furniture looked, she glowered at me and said "What are those antlers doing in the house!" "Remember my dear? That was our deal. I paid for the new furniture and you blessed the idea of hanging the moose antlers in the living room. Remember?"

I will never forget her words "I thought you were kidding about bringing the antlers in the house!" But a deal is a deal. She grudgingly relented. I had my moose antlers on the wall in the living room. And I had a new couch to sleep on…..!

The moose antlers are still on the wall. If anyone asks where that fantastic trophy rack came from, the answer is the same. "Got him over on Fotch Creek! He's a beaut, isn't he?"

Chapter 7

Emma Lee's Kudu

Every Boy Scout knows the Scout motto: Be Prepared. Not every Boy Scout knows that the founder of the Boy Scouts, Lord Baden-Powell, had a kudu horn that he used for summoning his scouts. It may seem strange the horn of a noble African antelope, a type used by the Metabele as a war horn in the 19th century, should call Scouts and Scouters together in the United Kingdom, America, and in many countries around the world. In the summer of 1907, Baden-Powell held his first experimental camp on Brownsea Island in Poole Harbor. Retrieved from his collection of African trophies, the kudu horn entered Scout service. It is integral, even today, to the BSA Wood Badge program.

Emma Lee's dad was a Scouter and had a kudu horn in the house. She remembers it from her childhood. When we decided to go to Africa for our "hunting-moon" in 2016, a kudu was the Number 1 trophy on Emma Lee's list. I procured for her a package hunt that included a Cape kudu.

John "Madala" Barnes scoured the hills and veld of the Eastern Cape Province looking for a suitable kudu bull for Em. I know, because I trudged along with them! While Em took a number of fine trophies, our hunt with African Cape Trophy Safaris ended without a kudu "in the salt" for Em. We saw some bulls at a distance. And we saw "scooter" bulls (young bulls with only two twists of horn and points that direct out like the handles of a scooter...) within range. But the only "shooter" within range showed only his head and neck above the scrub. Madala and Em decided together that it was too iffy a shot to take. We flew back to Johannesburg without her Number 1 trophy on her wish list. There were too many other neat things going on for her to be terribly concerned about this development, but I could tell she was disappointed.

We went on north to Swartwater to the Sekombo Lodge in the Limpopo Province. Em had been booked on this hunt as a "non-hunter" accompanying me while I chased Cape buffalo and perhaps another bushbuck. But as the safari evolved, an alternate plan evolved...

Emma Lee's Kudu

Before we left for Africa, we all became proficient at shooting "off the sticks."

Hunt Day One dawns clear and 50 degrees. A perfect hunting day. We watch waterbuck and bushbuck along the Limpopo River bed while we eat breakfast! And there are hippo tracks within 30 yards of our table. Em stays in camp to rest up from the travel day yesterday. Steyn and I will be hunting bushbuck today. Still-hunting our way through the thick riparian brush. Much like still-hunting for whitetail deer as I am accustomed to doing back in the USA. We head out, stopping briefly to collect Steyn's head tracker Patrick. Before we have covered too many miles, we are seeing kudu and kudu sign in abundance. A thought crosses my mind and I inquire of Steyn. Does he have plenty of kudu on the property? Any chance we could alter our hunt contract and let Em try for one of those kudu? Happily, Steyn agrees to let Em hunt kudu for just the trophy fee if she is successful. Chasing kudu will need to wait until after Steyn and I return from our side trip south for Cape buffalo, which is the main event on this safari.

Emma Lee's Kudu

Steyn and I return to Sekombo from Mpumalanga after our successful buffalo safari. After stopping at Polokwane for a few groceries, we head on north to the camp. We make one more stop, this time at a roadside food stand near Baltimore for what might possibly be the absolutely best grilled sandwich I have ever sunk my fangs into! I try ordering the Junky Monkey (cheese, sausage and tomato) but alas one ingredient or another is missing or in short supply. I settle for one of finely ground chicken salad that is utterly fantastic in its own right. Back at camp we celebrate with a stir fry of strips of wildebeest backstrap and fresh garden vegetables. Dessert is apple koek with vanilla ice cream. I am forced to conclude I will not be losing any weight on this trip! After showering this evening, I come to realize that I have picked up some "hitch-hikers" while down in Mpumalanga. On my lower legs, I have about 50 bites from "pepper ticks." These little rascals are no larger than of a grain of pepper. Hence the name. But the aggravation and itching they bring is way out of proportion to their size! They are the South African equivalent of the chiggers of the American south. Not fun to suffer through. And only time will solve the issues. I finally get some sleep after applying some cortisone lotion.

My first morning back at Sekombo after the buffalo side-trip to Mpumalanga is clear and 53 degrees. I'm still hunting in shorts (probably the main reason the ticks found me an easy target...) It's crisp as we leave camp in the morning, but by 10 am, I will be the smart one. We partake of the Afrikaaner version of "French" toast and thick slices of bacon. One last cup of coffee and off we go again to still-hunt for bushbuck. We spot a fair number of bushbuck, but only a couple of rams. Nothing of note shows itself. Patrick walks back to retrieve the bakkie, which is parked at Hippo Gate. Em, Steyn, and I make our way down river to the lodge. When Patrick gets back, we load up and go check some trail cams. Our original plan had been to get Em into a ground blind for kudu right after lunch. But her bad back is bugging her. We put this plan on hold; Em will rest for a while at the camp.

I'm relaxing and reading in the shade of the lappa when Steyn suddenly proclaims: "There's a pretty good bushbuck ram out here. Would you like to try for it?" Do pepper tick bites itch? Hell, yes I would! When I pictured

Emma Lee's Kudu

Emma Lee risking her very life as she enjoys the sights, sounds and smells of an African "game park."

myself taking a trophy bushbuck, the vision was of stalking through the thick riparian brush and shooting off-hand. Instead what I have is my butt in a canvas chair and a solid rest on the railing in front of the braai pit! One adapts as one needs to. Steyn hands me the 7x57 from out of the rack on the bakkie. We range the ram at 160 yards. I squeeze off a rock solid shot and the ram goes only a few yards. He is a fine specimen. An old, old male. Dark colored, with mud-encrusted horns. Steyn whistles up a couple of the staff who trot out and fetch the ram. We take pictures out on a sandbar in the Limpopo, in full sunlight. The beast is hauled off to the skinning shed and we get back to concentrating on consuming lunch.

I try and convince Em to go out and hunt this afternoon, but she would rather rest. At about 4:30, Steyn and I check trail cams again. Interestingly, the kudu did not drink at their regular time and place today. This means that Em unwittingly made the right decision to stay back and rest. She didn't miss anything!

Emma Lee's Kudu

Lions roaming on private land are either visitors from nearby national parks or, if behind fences, regulated by a ton of fencing requirements.

Supper consists of grilled pork steak, green salad, squash (mashed pumpkin) and cold, shredded cabbage with raisins and pineapple. Dessert is yet another Afrikaner koek, this time adorned with ice cream and some sort of custard. We retire to the campfire after supper and then to our room just before 9 pm.

My tick bites are acting up; I am fairly miserable. Moisture and heat seem to aggravate them. After my shower, I use the ointments Steyn has provided from his med-kit. I dab each bite with the South African version of bacitracin to ward off infection and follow up with more cortisone cream for the itching. I am very much feeling like I should have brought some Benadryl with me from Alaska. It takes a long time to get to sleep.

Day Six features yet another fine day in the African veld. Other hunters arrive. It sure has been nice to have been the only clients in camp, but that honeymoon is now over. The new visitors head out to pursue impala.

Emma Lee's Kudu

Steyn and I investigate some crocodile tracks coming out of the river and going into a stock pond behind the camp. Checking trail cams, we observe there are a couple of shooter kudu back on their regular schedule. Things are looking up for Em. There is one minor problem; Em's PH is about to bail on her.

Steyn has a line on a potential croc hunter, but there is a fair amount of paperwork involved in making this plan come together. Were the croc to stay in the river, it would not be legal game. Once he leaves the river and takes up residence in the stock pond, he is potentially classified as a "problem animal" and, with the proper paperwork, could be taken. First Steyn needs a permit from "Nature Conservation" in Polokwane. He must apply there in person. It seems that our kudu quest might be on hold. As it turns out, Steyn, like every good professional hunter, has a plan.

He sits me down at the table in the lappa and says: "You are a professional hunter in Alaska, right? You have hunted kudu a few times, right?" I reply in the affirmative to both queries, but must ask "What does that have to do with Emma Lee's kudu hunt?"

As it turns out, Steyn is in a yank to go to Polokwane. He says out loud what he is thinking. If he furnishes me with a bakkie and Patrick the tracker, in his plan I should be fine as a newly minted "junior chipmunk" South African PH. I am not as convinced as he is, but I'm willing to give it a try. For some reason, Em trusts me to do it! This should be fascinating. I will perform in this regard for my first outing while Steyn is still around. Once he leaves for Polokwane, I will be on my own.

Susan packs us a picnic lunch. Steyn drops us off at the water hole. We begin to munch the tasty goodies, but the Zoo Parade starts before we get very far into Susan's goody basket. Waterbuck, zebra, impala and a few blue wildebeest. It's getting close to sundown. I've gotten a ton of photos, but no kudu has shown itself. We begin packing to be ready to be picked up. While I am otherwise occupied, a kudu bull materializes out of the dusk. This bull is young and nowhere near being a shooter. We watch him until darkness falls. Steyn pulls up in the bakkie to collect us. We load up and drive back to the camp.

Emma Lee's Kudu

This is the handmade blind from which Emma Lee shot her kudu.

Back in the kitchen, Susan has prepared another Dutch oven masterpiece.... lamb potjie. And yet another koek.... Tangerine Naarje. Said Tangerine Naarje koek is so good I can hardly stand it. It takes me three slices to be sure of how good it really is!

The next morning is Friday, May 13th. Clear and 50 degrees here on the banks of the Limpopo. Steyn is doing his happy dance. The croc is back in the stock pond and the hunter is on call. Now he just needs the "problem animal" paperwork and a CITES permit. All hands on deck to protect people, game and livestock... and make a few Rand in the process! About mid-morning, things are moving rapidly. Steyn wants Em and I in the blind right away so he can take off for Polokwane. Em thought she had another hour and a half to get ready. Her schedule is now in a complete dither.

Within 20 minutes we are back in sync. Steyn has off-loaded the alfalfa hay procured earlier this morning and is taking his chances "helping" Susan and

Emma Lee's Kudu

Gladys in the kitchen. Em is out of her night clothes and into her camo. Susan has packed a terrific lunch. Patrick is standing by with the bakkie. What could go wrong? Even on Friday the 13th?

What goes wrong is that Steyn is fairly anxious to get to Polokwane. He takes off like a NASCAR racer to drop us at the blind at the water hole. The rushed ride results in bade vibes for Em's back. She arrives at the blind in less than a perfect sense of humor. The first 15 minutes in the blind are spent listening to Em lecturing me about Steyn's driving habits. "Suck it up, Buttercup. Unless a kudu comes in early, you will be here in this blind for the next six hours!"

We partake of our picnic lunch. Grilled tuna melt sandwiches, leftover koek, bananas and tangerines. A Coke Zero for Em and a Ginger Beer for the up-and-coming novice PH from Alaska. The water hole remains completely deserted for quite some time. A weather front is forecast to move in; the associated swirling winds are not particularly helpful. It is almost two hours before the first animals show themselves. First, a huge male waterbuck, followed by a parade of young impala, guinea fowl and more waterbuck. Finally, around 5:20 pm a young "scooter" kudu bull drifts in for a drink.

Five minutes after the young bull moves in, he is joined by another bull, this one from the southeast. I slowly raise my binoculars and quickly realize that this is the bull we have been waiting for. Not only is it a shooter, it is a VERY good bull. We have no time to waste; daylight is fading fast. I get Em up on the shooting sticks. She keeps hissing at me to "back up and get out of the way!" My concern is that I need to be close enough to whisper instructions without spooking either bull. Em is ready, but now there is another problem. The two bulls are completely lined up. She cannot shoot without chancing a pass-through that might wound the younger bull. We ponder this predicament, but it solves itself. The young bull moves out of the way, leaving the big bull in the clear.

Emma Lee is more than ready. She is on the sticks, finger along the trigger guard and the big bull's shoulder in the scope on its lowest power. The bull is quartering sharply toward us. I advise aiming for the "point" (juncture of

Emma Lee's Kudu

Emma Lee with her magnificent Cape Kudu bull, the first (and last) kudu that Pete ever guided for!

the scapula and humerus) of the left shoulder. This advice points out my failings as a guide. Apparently we have not discussed enough anatomy for this to be a clearly understood set of instructions. The bull solves the problem by abruptly turning full broadside. Em now quickly and correctly adjusts her point of aim to a spot just anterior of the point of the bull's left elbow.

"Take him" is the next bit of solid guiding advice Em hears from me. The 7x57 barks. Obviously well-hit, the bull humps and lunges forward a few yards. Em can see him, but we had agreed I would back her up as necessary with the .375 H&H. I exit the blind to the rear. I can see a bull kudu standing in the thorn-bush, but I cannot make out for sure which bull it is. I cannot chance a backup shot. This bull sees me and charges off. But in a few seconds I hear him crash to the ground. I haul out my tiny flashlight; it is not strong enough to help much.

Emma Lee's Kudu

What to do? I start by calling for reinforcements on the radio. Per the plan, Patrick should be parked fairly close by in the bakkie. "Patrick, Patrick. Kom in..." Patrick answers my radio transmission immediately. "Emma Lee has shot a kudu. Bring the bakkie!" "In a moment..." he replies. In less than five minutes, Patrick pulls up behind the blind.

Emma Lee is sure of her shot, but as with any hunting situation where the animal does not drop immediately, there is that tiny nagging doubt until you actually put your hands on the quarry. Nightfall is coming on quickly. Em stays in the blind while Patrick and I don our headlamps and take up the track. We don't have to go far. Within 20 yards, we find the bull. He is "graveyard dead!" We call Em up to us and the three of us celebrate. No black or white here; just three hunters!

It is a magic moment. Em runs her hands over the magnificent bull's mane and up along the chevroned face and spiral horns. She has waited for this moment for a long time. The photos reflect the magic. With Em operating the winch on the front bumper of the bakkie and Patrick and I wrestling the carcass, slowly the kudu comes on board. On the way back to camp, we meet Evert coming out. Bless his heart for checking on us. I suspect what was really going through his head was: "I've got to check on the crazy Alaskans who disappeared while hunting kudu, while the real PH was in Polokwane visiting his girlfriend and obtaining permits to hunt a croc!"

Friday the 13th certainly turned out to be Emma Lee's lucky day. We feel very blessed by the hunting gods.

I'm reminded of Em's words as we left our house in Fairbanks. "We have enough furry things on the wall. No more taxidermy mounts." Now she is singing a radically different tune: "Perhaps we can find a contractor to remodel the living room. We should raise the ceiling so as to properly display my kudu shoulder mount!"

Chapter 8

Arctic Death March

Author's Note: There are few photos to document this adventure. I can explain why..... my camera froze! You're going to have to imagine stuff...!

One of the more fascinating of Alaska's many species of big game is a reintroduced species, at one time extirpated from our state. Muskox had once been a completely circumpolar species, but by 1920 they were gone from Alaska and much of their traditional range in some other countries. They survived in significant numbers only in Greenland and arctic Canada. About three dozen of the animals were brought to Alaska from Greenland in 1930. They were moved to Nunivak Island in the Bering Sea. Here they flourished and subsequent relocations were made from the Greenland bloodline on Nunivak to several other Alaskan locations, including the North Slope of the Brooks Range, east of the Haul Road and south of the village of Kaktovik, on Barter Island.

By the 1980's muskox were spread across much of the North Slope. Some had wandered east into the Canadian Arctic. One was even spotted on the Sheenjek River on the south side of the Brooks Range in Alaska. They were very common along the Haul Road as far south as Slope Mountain. ADFG decided there were enough animals and the herd was growing steadily. It could sustain a limited hunt. A few permits would be available in Kaktovik and a few in Fairbanks. It was "first come, first served," so I stood in line (camped out) at the ADFG Regional Office in Fairbanks and ended up obtaining a coveted bull muskox permit, as did my friend Lynn. What could possibly go wrong?

The first inkling that something was amiss, was when my hunting partner (who had camped in line at ADF&G with me) punked out on the hunt. I decided to charge on by myself. I lined up a flight on the mail plane. I found an enterprising young man in Kaktovik who agreed to rent me a snowmachine "in top condition," as well a qamutiik (a long wooden sled pronounced com-a-tic by Caucasian lips that dare to try...) as well as a wall-

Arctic Death March

Spring in Kaktovik, on Barter Island, is still brutally cold, with many visual reminders of just how cold it was back in Winter!

tent and a vintage Coleman stove. I got permission to sleep in a bunk house in Kaktovik commonly used in summer by geologists doing oil exploration work. Things were coming together.

I hopped on the mail plane and flew to Kaktovik on March 20. It was still winter in Fairbanks. It was "industrial strength winter" at Kaktovik on Barter Island in the Beaufort Sea. Serious cold and serious wind. I spent a couple of days laying low as I waited for the sideways snowstorm to break. Break it did and off I went. Bright sunshine, and a brisk -30 degree Fahrenheit (F). Rough traveling, but on the bright side, if I got a muskox, meat spoilage likely wouldn't be a huge problem.

One nice thing about winter travel at this latitude is that with so much wind, the snow for the most part becomes windswept and packs down quite firmly.

Arctic Death March

Actually it becomes as hard as a rock. I motored off the island, across the ice of the Beaufort Sea and onto the mainland. The qamutiik pulled nicely on a long single line. The country was fairly flat and I made good time heading about 50 miles southeast of the village onto the heart of the Coastal Plain of the Arctic National Wildlife Refuge. Soon I was seeing muskox sign. This was probably going to be easy. Right........! Sure it was!

By early afternoon I had seen a lot of sign but exactly zero muskox. When I came to an old oil exploration camp with a metal building I said to myself "Self, it's going to be a lot less windy in that metal building than it is out here in a drafty wall tent." I went in and made camp inside the building. I had some soup. Later I boiled some water for tea and had a freeze-dried meal. I read by candlelight for a bit and then slept well for the entire night. In those hallowed days of my youth, I could go all night long without getting up to pee. That logistical consideration was a blessing indeed in the high Arctic at 30 degrees below zero!

At first light I was out of the sleeping bag and up on a tower outside to glass for muskox. And there they were... Within a mile and a half of the camp, a small herd of muskox was pawing through the drifts, munching on grasses and sedges. I couldn't see them all the time, due to blowing snow. In retrospect, that should have been a clue that I should have paid attention to!

I got the Skidoo started; no mean feat in the chilly temperature. Then I motored a bit closer to the herd before stashing the machine in a depression in the frozen tundra. I walked several hundred more yards concealed by terrain. When I popped up, I was within 50 yards of the small herd of muskox. One would think that at 50 yards, one could clearly consider the situation and make good decisions. If you were thinking that, you would have been wrong in this case!

My permit was for a bull muskox. I had been through an orientation course to tell the boy muskox from the girl muskox. I had spent time actually laying my hands on live muskox at the University of Alaska's Large Animal Research Station outside of Fairbanks. I have a degree in Biology and a Registered Guide license in my pocket. You would have to assume I could tell a bull muskox from a cow. But alas, when I pulled the trigger, I proved beyond a shadow of a doubt I could not!

Allow me to mount a feeble defense... It was -25 degrees F. It was snowing and the wind was blowing at least 30 mph. Muskox have long hair, as long as 18 inches in places. Seeing the normal anatomical "proof of sex" (that is to say, external plumbing appendages) is problematic at best. Cow muskox, like bulls, have horns. The significant difference is that on the cows, the horn bases do not meet in the middle of the forehead. Bulls have a great wide boss. Cows just have some white fuzz between the horn bases. The animal I had in my sights had really long horns (for a cow...!) I pulled the trigger and the animal went down.

I moved up to the fallen muskox. I still thought I had killed a bull. After a quick photo or two, I rolled it over to begin the field dressing. Imagine my surprise when I parted the long hair in the beast's nether regions. In hunting parlance, the plumbing fixtures I found were more in the line of a "4-pointer" (on the udder) instead of the "spike" I was expecting. Parting more hair and examining more plumbing did not prove encouraging. I had mistakenly killed a cow muskox. Oops. Time for what our wildlife troopers euphemistically refer to as a "self turn-in." But communication from where I was standing in a ground-blizzard on the North Slope was out of the question. In no way a viable option.

I completed the field dressing. I fetched the snow machine and qamutiik and rolled what had been my prize and was now probably just "State's evidence" onto it. I motored back to the metal building, quickly packed my gear and loaded it on the sled with the carcass. It was now mid-afternoon. The smart thing to have done was to get into the metal building and spend the night. But I am, ahem, somewhat famous for not always doing the right thing.

I figured I could rapidly travel back across the ice of the Beaufort Sea, following the coast. I had about four hours of daylight left. Even after sundown, I should be able to see the strobe lights at the airport and the DEW (Distant Early Warning) Line Station at Kaktovik.

Arctic Death March

I had now used up my four hours of remaining daylight. I made it to just east of the mouth of the Jago River before the poop hit the prop. Suddenly the Skidoo made a very expensive noise and came to a halt. I tried to restart it and could not. In fact, the engine would not even turn over. Completely seized up. My trusty arctic steed had blown a cylinder. All I could think of was the many times I had advised "lesser" outdoorsmen never to venture out alone on a single snow machine. 'Twould have been some good advice to heed myself!

But here I was. Alone, on the ice of the Arctic Ocean, in the dark, at 30 degrees below zero, with the wind blowing 30 mph. Ol' boy, you find yourself in quite a pickle, don't you? And it's nobody's fault but your own. Now what? You better think of something clever and think of it fast!

It was not like anyone would come looking for me for at least a couple more days. I was not expected back in the village until then. I could wait out the storm if I could put up the tent and get the Coleman stove going. But, in between gusts of blowing snow, I could actually see the strobe lights at Kaktovik. It was probably the Long Range Radar domes at the airport. If the wind stayed steady, I could walk into it and come to Barter Island. At issue was the fact I would have to walk INTO the wind. That detail eventually became an issue of its own.

For some dumb reason, I decided to try and walk to Kaktovik. It was not my most brilliant decision ever. In fact, one of my all-time worsts. It could have easily cost me my life. I put together a small pack of survival gear and set off toward the strobe lights.

I trudged into that fierce arctic breeze for almost six hours before wearily walking onto the end of the Kaktovik runway and making my way into town. I stumbled into the structure I was staying in and fell exhausted into bed. The blizzard still raged outside. I had survived my self-administered overdose of "stupid."

The storm blew itself out in a couple of days. The owner of the decrepit snow machine and his friend went back over to the Jago River with two other

Arctic Death March

machines and a second qamutiik to retrieve the deceased Skidoo and the now solidly frozen deceased muskox. Both were brought in and stashed in a heated structure to thaw. It took only a day to replace the cylinder in the machine. It was four days before the muskox thawed enough to be skinned and cut up.

I occupied part of my remaining time in the village by calling the Alaska Wildlife Troopers and confessing to have accidentally killed a cow muskox while holding a bull permit. They told me to come into the office when I got back to town and to bring my muskox tag. I was instructed to leave the meat with a North Slope Borough Public Safety representative for distribution to needy local families. It all sounded innocuous enough and a good resolution to a thorny problem relative to breaking a wildlife law.

It was only when I got back to Fairbanks I learned on the very day I had shot the wrong muskox, another pretty wild event had occurred in Alaska. The Exxon Valdez had run aground on Bligh Reef and leaked a lot of oil into Prince William Sound. I always meant to contact Captain Hazelwood and thank him for being the shit magnet that diverted the world's attention away from my transgression of March 24, 1989. (P.S.: I was never arrested for my alleged crimes...)

Chapter 9

Mentors

Finding a hunting mentor can be a daunting task. My dear old dad, bless his heart, was a great dad, but he was born and raised in New York City. Although he adapted well to a rural lifestyle when he became smitten with my mom and moved full time to the country, he was not a hunter. He was a gentleman and very civilized. But he was in fact, the antithesis of a country boy.

My mom however, came from a rural/agricultural background. Going back many generations, her side of the family had hunted, fished and farmed. There were some bankers and judges in the line, but even they were ardent hunters and fishermen. The upshot of all this genealogy was that it was my mom who got me started shooting and hunting. Mom graduated from Cornell University School of Veterinary Medicine at Ithaca, New York (NY) in 1942. This was long before it was fashionable for ladies to become vets.

Mom would come to Alaska each winter for a month or so. She worked as a Trail Vet on the Yukon Quest International Sled Dog Race between Fairbanks and Whitehorse, Yukon Territory. Once, in the 1980's, a Fairbanks Daily News-Miner reporter inquired of my mom during an interview: "Dr. Buist, you were pretty much a pioneer in your field. What do you think about the Women's Liberation Movement?" Mom answered with characteristic bluntness: "Sweetie, there are those that talk about it and those that DO it!"

When I was younger and it came time for me to start hunting, Mom was a tad busy. She taught me the basics of shooting and ensured that I went through Hunter Education before obtaining a hunting license. But I needed someone to teach me the next levels. Enter Donald C. D'Amato.

I first met Don while I worked as a camp counsellor at the New Jersey School of the Outdoors in Stokes State Forest. In "real life" Don was a Seventh Grade science teacher in the Newton, NJ school system. He and another teacher/

Mentors

principal from the Branchville schools, Bob Haight, were working at the School of the Outdoors. These educators were terrific mentors. Both were graduates of forestry schools, Don at Syracuse and Bob at the University of Maine. I had a 4-H forestry project when I was in 7th Grade; I was pretty sure I wanted to pursue forestry as a career. Listening to these guys only fueled the fire even more to start a career in forestry. Don had some added attractions going for him. He was a hunter. He harvested a lot of the protein that his family consumed. Deer, ducks, geese, grouse and other common small game. Fish and even eels from the weir on the Paulenskil River!

Don was also very good at producing extremely good-looking female progeny! From his first marriage he had three daughters, Ann, Donna and Nancy. In due time, the second marriage produced Lynn and Lauren. From the initial classification of the older girls as targets of opportunity.... these ladies became partners in crime and my "sisters from another mother." They remain so to this day.

Don knew more about deer hunting than any person I have ever known. He lived in a very rural portion of Warren County, NJ. We hunted deer out his back door. From the beginning of the hunting season to the end, I would rush to his house every day as soon as school was out. We would comb the woods and fields for deer and small game. Don taught me to take proper care of the game we took. And he taught me how to cook wild game. He was damned good at it and I was an attentive student. Don's cooking ability did not make me fat, but it kept me hanging around. I ended up earning a Forestry degree at Don's alma mater, the State University of New York, College of Environmental Science and Forestry at Syracuse. And soon after I graduated, I was married. Don was my Best Man. He likely was relieved that I was no longer lusting after his daughters.

Right out of college, I was drafted into the Army and ended up stationed in Alaska. Don loved sun and sand. He retired from teaching in NJ and moved to Key West Florida for a spell. Then, as he put it, "moved up north," first to the Fort Myers area and later to Gainesville, Florida. I owe the hunting bug in my blood to Don. He had taken me hunting hundreds of times. In 1979, while I working as a guide in western Alaska, I invited him up for a

Mentors

My Mom, the late Dr. Jean Buist, loved all things outdoors, but especially her horses. She rode daily until she was 76; after that she drove.

caribou hunt after we were done with the regular clients for that season. Don had a ball. In fact, he shot a B&C caribou bull on that trip. The antlers took up residence beside the pool in Florida (FL) and were used as a rack for barbeque tools! Despite his penchant for sun, sand, the beach and warm temperatures, Don was hooked on Alaska. I found a mutually advantageous way to exploit that penchant.

It was just a couple of years later I began my own outfitting business. We all know that an army marches on its stomach. So does an outfitting business. I hired the best camp cook I knew and flew him to Alaska to cook for me. Don cooked for my staff and clients for a lot of hunts over many years. When the tent living got a little too chilly for the "Best Camp Cook on the North Slope," we hauled an old travel trailer with a propane heater in it to keep Don toasty and happy! He was a fixture in my North Slope camps and

Mentors

Red Beeman, of Chugiak, was a huge and positive influence on my life as a hunter and guide.

had a widely known reputation for producing "goodies" from the propane stove and oven that we carted back and forth to the North Slope. One time some resident hunters pulled into my camp at Galbraith Lake because they had heard up at Deadhorse, that there was guy in my camp who made the "best cinnamon rolls ever...!" When they left to drive south again, they were convinced it was indeed true.

Another extremely important positive influence in my life and on my hunting and guiding career was Edward "Red" Beeman of Chugiak, Alaska. I first met Red in 1975. Red took a chance on me and offered to hire me on as a "packer" (camp monkey, wood/water slave and general laborer) at his lovely camp on the North Fork of Big River in GMU 19. I recall hopping off the charter flight over the Alaska Range from Anchorage and landing at "North Fork." Within minutes, Red was tying a Coot (articulated 4WD

Mentors

off-road vehicle) engine onto my pack board and pointing in a northerly direction. "Pack this Coot engine six miles out this trail. Cover it with a tarp. Try and be back by dinner!" I did it and so passed Red's first test.

I loved my time with Red. Red was an "old time" guide and the real deal. He was fond of saying he learned woodsmanship from Talkeetna Mountains pioneer guide and trapper, Oscar Vogel. He learned "guide showmanship" from Rainy Pass Master Guide Bud Branham. I got the best of both those educations, all rolled into one, from Red. I worked several years for Red and became close to his family, including his wife Bunny, son Eric and daughter Sue. He really was my "Alaska Dad." I was and remain so very grateful to him for teaching me about cabin building, carpentry and of course the hunting and guiding business.

After serving a couple terms on the Alaska Guide Licensing and Control Board and later the Big Game Commercial Services Board, I observed that young assistant guides really reflect not only the abilities, but also the ethics of the guides for whom they apprentice. I saw some of the old time airborne outlaws lose their licenses. Some went to jail. Soon many of the guides who apprenticed under them were all grown up and losing their own licenses! I became doubly grateful that my guiding and outfitting ethics had been imparted to me by Red Beeman. And I'm glad I was paying attention most of the time! I later served a term on the Alaska Board of Game and now I am back on the Big Game Commercial Services Board. Not a session goes by while I'm sitting on those regulatory boards getting ready to vote, that I don't find myself thinking, how would Red handle this? It is comforting…

I would be remiss if I did not mention two Athabascan elders, now both deceased, who were mentors for other facets of my life. The first was Howard Luke. When I first began trapping the Tanana Flats, south of Fairbanks, I inadvertently ended up encroaching on Howard's trapline. I was horrified, but Howard not only forgave me for my transgression, he basically "gave" me a bunch of the trapping real estate he had earned and established as "his" line. As if that were not enough, he taught me more about mink behavior and mink trapping than I deserved to learn. As the years went by, I spent a lot of time at Howard's allotment and camp on the Tanana River. He taught

Mentors

me how to build, site and run a traditional Athabascan fish wheel. Howard died in 2019, at the age of 96. I am indebted to this kind man who took me under his wing, shared his culture and taught me so much.

My other Alaska Native mentor was Sidney Huntington of Galena, Alaska. Sidney's life is immortalized in the book Shadows on the Koyukuk which he co-authored with fellow Board of Game member the late Jim Rearden. Sidney donated his time and talents to the people of Alaska in many ways. In my world, this was exemplified by his service for over 20 years on the Alaska Board of Game. This incredible individual brought an intricate knowledge and understanding of life in rural Alaska, primarily the Yukon-Koyukuk region of his birth, to his public service, including the Board of Game. Some of the finest entertainment I have ever experienced was watching Sidney debate the merits and biology of predator control with elite, rich representatives of Outside animal rights groups testifying before the Board of Game. A particularly notable example was his response to an activist lady from back East who was taking questions from board members. After her indignant and unhinged rant, the chairman called on Sidney. The question he posed to the arrogant lady was this… "Ma'am, I have just one question for you. Have you ever been so hungry that when the moose were gone you were reduced to making soup out of ptarmigan shit just to survive? I have…" I hasten to add that there is apparently no particularly good response to this question! Sidney had made his point in spades. Years later, when the Governor appointed me to the Board of Game, the very first person to call and congratulate me was Sidney. With the congratulations came a few helpful suggestions on how to comport myself in the job! I paid rapt attention and the advice served me in good stead during my time on that board.

Through my appearances and testimony before the Board of Game and my work as President of the Alaska Trappers Association, I came to know Sidney quite well. For some reason he took a shine to me. We became fast friends. Actually it was more a case of me, a low-level apprentice monk, sitting at the feet of the Dalai Lama! Lord knows I paid attention to the pearls of wisdom that were offered. Later, I was overjoyed when Sidney eventually came to ME seeking help. He wanted to establish a commercial fish smoking business at his camp on Daisy Island in the Yukon River near Galena. He had been getting the bureaucratic runaround from folks at the Bureau of Land Management (BLM.) I spent many an hour talking with folks at BLM on Sidney's behalf. As well as more time relaying and interpreting the information I had gleaned. Sidney eventually got his lease.

In 1989 Sidney was awarded an honorary PhD from the University of Alaska - Fairbanks. I was pleased and honored to be present at graduation and cheer as he received this honor. I also watched as Sidney was presented the Alaska Trappers Association "Trapper of the Year" award in the Spring of 1988. In this case, he took possession of a large plaque with an Alaskan #9 trap bolted to it. In typical Sidney fashion, he unbolted the trap from the plaque, rendered it scent-free, set it out in the woods across the Yukon River from Galena near Yuki Slough... and promptly caught a wolf with it! Classic, and not at all out of character. Sidney passed away in 2015; he was 100 years old. I am honored to have been his friend and I miss him a lot.

Since I had fine mentors, I decided a long time ago that if I could, I would try and serve as a mentor myself to return the favor. I was lucky to have had a fantastic son, Jason. Born in 1976, Jason grew up being exposed to my outdoor obsessions as well as my eccentricities. He survived nonetheless. As a kid in Alaska Jason took moose and bear and did a ton of small game hunting. After getting a degree in Fire Science from the University of Alaska – Fairbanks, Jason went to work for the City of Kodiak as a firefighter. After a year toiling for the city, he moved over to a civilian firefighter position at the Coast Guard Base in Kodiak. Working on Kodiak Island gives a guy a chance to work in close proximity to a plethora of outdoor activities. The fishing and hunting are literally in your lap and just minutes from the house. Jason immersed himself in all this. He fished for salmon and hunted deer. And he began his lifelong love affair with hunting migratory waterfowl by pursuing the incredible array of sea ducks that spend time on Kodiak Island. He became enamored with shotguns. My only regret? At first Jason was not as captivated by big-game hunting. He sure made up for it with his intense love for duck and goose hunting. Later he turned into a pretty skookum big game aficionado and a dedicated turkey hunter. I'm OK with that gradual metamorphosis!

Mentors

Jason tagged along as a packer on this Brooks Range sheep hunt in the mid-1990's. Always good to have a stout young lad along as a pack horse.

If having one kid to mentor is fun, it's even more fun to have more! Toward that end, I found myself happy to help provide guidance and opportunity to another fine young man. Peyton Merideth, I'm fond of saying, is "not the fruit of my loins, but he sure is the fruit of my refrigerator!" Jason and Peyton met at a very young age in a local youth group and became pretty inseparable (with a brief time out when they discovered girls...) In a variety of settings I am addressed by Peyton as "Dad #2," which is indeed a special honor for me. Peyton too attended the University of Alaska - Fairbanks, earning his degree in Criminal Justice in short order. He went on to become a commissioned law enforcement officer, working first at North Pole Police Department and later rising to the rank of Lieutenant at the Fairbanks Police Department. In 2019, Peyton retired from FPD. He moved his family to Idaho and took a job as an Investigator with the Ada County Sheriff's Department where his new hobby is applying for permits for the hunting of lots of big game species not found in Alaska.

Mentors

Jason and Peyton helped out in bear camp and then got to shoot bears themselves after we were done with the clients.

Peyton and Jason hunted, fished and trapped with me as they grew up. In their late teens they both worked for me in my outfitting business, doing their time as "packers." Once I had them all trained up and pointed in what many would agree would be the right direction, I had a few years off. The boys got married, started families and pursued hunting activities on their own. I sat with Peyton and "Little Peyton" one spring evening when Little Pey shot his first bear. While skinning the bear, Little Pey announced: "Uncle Pete, you helped my dad get his first bear. And you helped me get my first bear. If you are still alive then, wouldn't it be fun if you helped MY son get his first bear?" "Out of the mouths of babes…!"

The temporary "kid void" was soon filled and not just with grandchildren. I was approached by Cathie Harms, a friend and long term employee at the Alaska Department of Fish and Game. Cathie worked with a number of educational programs at ADFG, including a summer Conservation Camp. From the ranks of, first the students, and later the counselors at Conservation Camp, Cathie came up with two young men who wanted to become more

Mentors

Peyton took this bull muskox on Nunivak Island. I was there in spirit; he was wearing my fancy red parka!

involved in hunting, but whose home situations were not particularly conducive to participation in these activities. Thus I got to work my wiles on Harry and Calvyn. These guys and I shared the great Alaska outdoors over the course of several years, including hunting for moose and bear. Calvyn and Harry learned a few tricks of the trade and to hunt safely. They are now hunters in their own right. I like to think I was a pretty good influence for the most part. I stay in contact with these fine young men to this day.

I still do some teaching as an Alaska Hunter Education Instructor, primarily teaching Muzzleloader classes. I also present some trapping classes for the Alaska Trappers Association. I developed a training session called "Sharing Alaska's Trails" to mediate and referee conflicts between trappers and other trail users. I sometimes teach at ADFG's Becoming an Outdoors Woman clinics. Thus I keep my hand in. Plus, I have this army of stout young lads (and now some granddaughters) on whom I can call if the "old man" needs help packing moose meat, skinning a bear, or moving a gun safe or a fridge! What goes around truly comes around and for that I am grateful.

Chapter 10

So You Think You Want To Be A Guide?

In my early twenties, I was pretty darned sure there was not a lot I didn't already know about hunting. To be sure, I had to re-learn a lot of what I had learned in NJ, but my basic skills served me in good stead. Moose were certainly not smarter than a farm country whitetail. I learned I could use my deer hunting skills to become pretty dang good at moose hunting. Once a moose was down, my butchering skills, honed at slaughtering time in the back East barnyard, served me in good stead as well.

Then I discovered that a guy could actually be PAID to go hunting! Oh, for the life of a guide! This was too good to be true. This I gotta try...

In the spring of 1975 some Army buddies and I went on a do-it-yourself (DIY) brown bear hunt on the Alaska Peninsula at the western end of GMU 9. Four of us drove to Anchorage, stayed overnight at our former battalion commander's quarters at Ft. Richardson, then flew out of Anchorage to Cold Bay on Reeve Aleutian Airways. From there we chartered back up the peninsula in an ancient Gruman Goose amphib plane and camped for three weeks. Three of us got nice brown bears.

In the next bay was a camp established by a real, live Alaska outfitter and guide. This fellow had seven clients and seven guides. At the end of the same three weeks, the guide camp had racked up just one dead bear to our three. One of the guides in that camp booked clients of his own for multi-species fall hunts. Red Beeman of Chugiak, AK, offered to hire any of us "rat-ass GI's" as packers and to train us as future Assistant Guides. We all indicated our interest. A couple of us actually went on to do that. I worked that fall for Red at his cozy camp on the North Fork of Big River in GMU 19. The $10 per day that he paid me probably paled in comparison to what it cost to feed a 6' 4" 185 lb packer. But boy howdy, did I receive an education. I nearly quit guiding before I even got that Assistant Guide license.

So You Think You Want To Be A Hunting Guide?

I met this old Coot vehicle in Red Beeman's camp. Several times it consumed its bronze drive gears and we packed new ones out and installed them.

Red's first clients that year had booked a 15 day, five species, mixed bag hunt. There was good news... they were from my alma mater, the great State of New Jersey. There was also bad news. These clients were brothers from an Italian contracting family in NJ. The Soprano brothers; before their time! Their social norms were, as it turned out, a bit different than ours. OK, a LOT different than ours.

Now granted, the weatherman did not do us any favors on this hunt. The clients arrived on September 5th during a break in some nasty weather. From that point on, it either rained or snowed (sometimes both) for nearly as long as these guys were in camp. We hunted first for sheep, trying to get rams before early snow drove us out of the mountains. Assistant Guide, Kelly, and I actually packed a wall tent, tent poles, food, cots and fuel as well as other camp items, 11 miles up the North Fork on our backs. Our New Jersey buddies would have no reason to be uncomfortable.

So You Think You Want To Be A Hunting Guide?

The wall tents in camp are on plywood platforms, with several rounds of logs around the base and windows and a door framed in.

A couple of days later, Red, Kelly and the brothers and I walked the 11 miles up to the new camp. We spotted a fair amount of game on the way up. The younger brother even shot a very nice caribou during the journey. Kelly and I would take a "day off" later to pack it all the way back down to main camp.

We hunted both clients out of this upper camp for several days. In fact they were there long enough they both missed fairly easy shots at sheep. A day or so later the older brother (the REALLY squirrely one...) shot a pretty darned nice grizzly. The weatherman was still heaping abuse on us. Conditions were somewhat trying, but not really gruesome by any stretch of the imagination. Fall hunting in Alaska very often is done under unpleasant weather conditions; it comes with the territory.

As the food box at the spike camp became a little thin, we hiked back down to main camp. This would be more pleasant. We would be in the timber and have warm, wood-heated wall tents. In fact, the tent platforms were darned nice.

So You Think You Want To Be A Hunting Guide?

Plywood platforms. Four or five rounds of logs and a nice wall tent pulled down over the attached frame. A framed in plywood door, wooden bunks with foam mattresses, glass windows framed into the walls and a nice pot-bellied wood stove. Warm and toasty. Better than spike camp by a mile, and a lot more "comfort" than I was accustomed to hunting out of under the best of conditions.

A couple of mornings later I was hailed by one of the clients. "Bring us some more wood" he ordered. I grabbed an armload of split, dry spruce and walked over. I knocked on the door and was invited in. I could see right off what the problem was. Both the damper on the stove pipe and the draft on the stove were wide open. The stove was consuming copious amounts of firewood and the heat was going up the chimney. I decided to show the city guys how to get more heat and conserve firewood. Sort of a "back woods win-win" deal.

I knelt by the stove, shoved some wood into it, closed the draft and stood up and closed the damper. Expecting something on the order of a "thank you," I stood up and turned back around. The older brother had picked up a loaded .44 Magnum revolver off the table. He shoved it up under my chin and looked me in the eye. "I said I wanted more wood" he said. "Not a lesson about stoves. Next time I need wood, how about I just shoot through your tent to let you know?"

My heart had pretty much stopped beating by now. With as much calm as I could muster, I told him I'd go get another armload of wood and bring it right back. And I did.

Then I went over to the cook tent and told Red what had transpired. He was not at all amused. When the brothers came over for breakfast, he had Kelly go pick up all the firearms out of the hunters' tent, wrap them in a tarp and hide them in the woods. When this was discovered, it was not at all well-received and our visitors got just a tad hot under the collar. Relations in camp continued to go gunnysack quite rapidly. My visions of a career as a guide were taking a serious nosedive as my dreams had a head on collision with reality!

So You Think You Want To Be A Hunting Guide?

Henny was my first bowhunter. He was a terrific hunter and a terrific client.

After breakfast Red and the clients had an interesting "come to Jesus" discussion about what was going to happen next. The hunters had enough cold, wet weather to last them for a while. They were ready to go back to New Jersey. Coincidentally, we had enough of their antics to last us for a while too. Red got on the HF radio and called an air taxi in McGrath to come get them. Later in the day the fog lifted and ultimately the wind dropped off. We sent the two brothers winging their way into McGrath. On the second aircraft we sent their gear and their firearms.

Later we would hear that they spent some time at McGuire's Tavern in McGrath, bad-mouthing Red. That did not earn our clients any new friends in the village. Red had trapped in the Farewell area for some time. His reputation in the upper Kuskokwim was already well-established. The tales the hunters told were taken with a grain of salt. Folks in McGrath will sell you booze and a meal, but that doesn't mean they will believe all your bullshit! The brothers caught the Wien jet flight the following day and headed back to Anchorage and on south.

So You Think You Want To Be A Hunting Guide?

Over at North Fork, we breathed a sigh of relief. We had three or four extra days before the next client was due in. I spent a fair amount of that time rethinking my idea of becoming a guide! Later in my career, I would proclaim that "I never met a client I could not tolerate for 10-14 days... as long as their check didn't bounce!" But after being held at gunpoint over a few sticks of firewood, I quizzed Red about how common this sort of thing was. Red of course explained that hunting clients were just like a cross-section of humanity. Some were idiots, some were fantastic and a lot were somewhere in the middle. I also remember some sage advice that came along with the suggestion that I hang with it for a bit and give it a fair trial. "You don't have to marry that guy, you just have to hang around him for a couple weeks!" Later in my career, I would dispense the same advice to young guides working for me and would relish doing so.

Our next hunter convinced me there really was a lot more to guiding than being held at gun point to obtain firewood. Henny was a gentleman and an accomplished bow hunter from Lititz, Pennsylvania. He had booked a 10 day bow hunt for grizzly bear. More to the point, he was the kind of client all guides covet and all clients should strive to be. He did not tell us how to hunt or where to look for grizzly bears. He happily ate the food we cooked. He was grateful and complimented the guides who cooked it. When it rained or snowed, he understood it was not the guides' fault. He was in better physical condition than most men his age. He was an accurate shooter, having with him one of the very first compound bows ever made, an Allen Brown Hunter, Model 7303. The only place he got "points off" from the packer was that he also brought along 16mm movie gear, including a huge, heavy movie camera and a huge heavy tripod! I got to lug that movie gear everywhere we went at North Fork. Everywhere...

Not surprisingly, Henny got a fine grizzly with his bow.

Our hunt with Henny restored my faith in humanity in general and guided hunters specifically. It confirmed that I did indeed want to get a guide license and try out the idea of getting "paid to hunt." Nearly 50 years and a lot of clients later, I'm sure glad that I gave guiding a good chance and stayed with it.

Chapter 11

Cowboy Mexico

At some point in time, I finally figured out that shooting a mule deer buck with antlers 30 inches or more wide in one of the western states in the U.S. was probably going to take more time than I had left roaming the earth. I had taken some nice bucks on some awesome hunts, but a 30 inch muley had eluded me. While hunting black-tails with my friends Doug, now deceased, and Janet in southwestern Oregon, I had discussed their mule deer operation in Mexico. They were taking some real toads out of the mountains of Sonora, near the little town of Carbo. Carbo was founded in 1888, when the main railroad in Sonora began construction. Carbo was one of the main stations established between the port of Guaymas and the border town of Nogales. Carbo began as a supply station for the merchants who came to the mines in the region of Rayon, Opodepe and Valle de San Miguel.

I succumbed to Doug's marketing spiel and gave him a deposit. I flew down in January of 2004 and hunted hard for a week without taking a "muy grande." The monsters were there. I saw a couple, but I never got a shot. I did get seriously hooked on the food prepared on the ancient wood cook stove in the *cocina* of that ranch house!

I was encouraged and booked again for a subsequent season. I have found that once you make an exploratory hunt in an area, you are much better prepared the next time. You know better the territory, the weather, the conditions and generally what to expect. I headed back south of the border two years later with high expectations.

January 2006 finds me flying from my winter quarters on the East Coast, to Dallas, then Phoenix and finally across the border to Hermosillo, Sonora. The flights themselves are uneventful, once I left the ground fog warning and tornado watch at Newark, NJ!

The front door to the hacienda at Rancho San Francisco.

In Phoenix I need to change terminals, but as I have checked my luggage all the way through, it's a matter of getting some exercise in the sunshine. My connection to Aero Mexico goes fairly smoothly. It only takes me two tries to fill out the Customs forms for El Presidente Vincente Fox. I'm struck how difficult the Mexican government makes it for Americans to cross the border and spend their money in Mexico! Meanwhile gobs of Mexicans are flooding north across the Rio Grande and wandering around in America.

Into Hermosillo on time. Since I will be borrowing a rifle and thus am not traveling with a firearm, I end up in the short line for Mexican Customs. Time slows down right as I am asked if I have a firearm: No. Followed by "Then why does the x-ray show that there is ammunition in your bag?" "Perhaps you are viewing an x-ray of someone else's bag?" "No Senor..." As it turns out the Mexican version of a TSA busy-body is looking at the X-ray image of a baggie full of AA batteries with a rubber band around it. Great work Senor Policia! One of his fellow officers points out the inconvenient truth to him. Several of the officers on duty have a good laugh at their compadre's

Handmade tortillas baking on the wood-fired cook stove.

expense. Meanwhile, this gringo is having a near heart attack and visions of the inside of a Mexican prison, where I seriously doubt there is even one 30" mule deer buck!

Soon I'm passed through to freedom with my duffel and my batteries. Janet and Luis are there to greet me, along with a client from Texas. We jump in the big truck and boogie off to El Walmart to fill up a couple of carts of groceries to haul to the ranch. As tradition demands, we stop at a street vendor hawking Mexican hot-dogs. We leave him with a few extra pesos, but we are richer by the hot dogs as well as some roasted green chilies stuffed with cheese, wrapped with bacon and grilled. When accompanied by ice cold Cokes, life is worth living again! The "inner man" is well-satisfied.

We arrive at Rancho San Francisco, rolling through the gate shortly after 7 pm. I reunite with Doug and Santi and meet another Texas client. After a quick supper, it's off to bed.

Cowboy Mexico

The next morning begins with my favorite alarm clock... Nidia banging around in the cocina fixing a pile of eggs, a mountain of bacon and a sturdy stack of French toast. Whatever happened to the *huevos rancheros*?

After breakfast Doug explains that the three hunting clients will rotate through *los tres guias*, Luis, Nito and Santi. Today I am to go with Nito. This is Nito's second year guiding; he was not here when I hunted in 2004. He appears to be about 45 years old and is said to own and work a small ranch nearby. Doug says Nito loves to leave the high-rack rig and walk in the desert. That suits me just fine. Our driver is Jesus. Off we go.

We happily hunt along, but see few deer this morning. At some point the fuel pump craps out on the truck, but after a quick SOS on the handheld radio, Santi and Manuel show up and haul us back to ranch headquarters.

Lunch is soup, frijoles and the ever-present handmade tortillas. More important, it is followed by the traditional *siesta*. I hunt with Santi during the afternoon. Supper is tacos and Nidia's homemade rice pudding. The deer situation notwithstanding, one should not intend to lose weight while participating in this adventure. After supper, Doug mentions there are a lot of *los chochis* (javelina) around this year. They are fair game and there is no trophy fee. In bed by 8:30 pm, anticipating a 5 am wakeup again. Too tired to shower.

Pete' del Norte's Spanish Phrase of the Day: Acaso huelo una enchilada vieja? (Do I smell like an old enchilada today?)

Up at 5 am. Turns out that MLK Day is not a big deal in Mexico; who knew? It is a brisk 46 degrees and clear. The busted down truck is home at ranch headquarters. Turns out the fuel pump malfunction was actually just a fuse for said part. The boys have replaced that fuse and the rig started right up.

I'm not sure what happened to the plan of switching the guides around, but I'm happy to be with Nito again. The only obvious drawback is that Nito speaks NO English and of course my Spanish is rudimentary at best. As it turns out, there is a secondary problem with being guided by Nito.

Cowboy Mexico

One night during the previous hunt, Nito cornered two skunks on the veranda and proceeded to beat both of them soundly to death with a stick. The smell was so bad for a bit that it made Olivia sick. Skunk Boy actually burned the clothes he had on that night, but the smell lingers on Nito himself!

We proceed down along the west side of Rancho San Francisco and onto El Pozo Hondo, the ranch to the south. We see a fair number of *burros* (mule deer) and a ton of quail. There are a couple of pretty fair bucks, but no *muy grande*. One big deer looks promising. We slip off the rig and follow him into the desert. With the temperature rising rapidly, the Nito-skunk interaction is recalled. We never see the big muley again, but we do spot a pretty fair *cola blanca* (Coues whitetail). At this point in time, I am happy to concentrate on finding a *muy grande burro. Cola blanca* are not on my to-do list at this point.

Back out to the main road on the Lariata side, just north of Enrique's place. Back through a lot of charcoal cutters' camps. There is a ton of mule deer sign here. I suspect the tops of the mesquite trees the cutters have felled and left are attractive to the deer. At one point we spot a band of does that simply must have a buck with them somewhere. We settle in and glass the area around the "girls" carefully. We are just 180 yards from them. Far enough they are not alarmed by our presence. Close enough that even I will be able to kill a big buck if he shows.

And soon enough, a great buck steps out. Nito quickly confirms my judgment that this is a "shooter" of at least 30 inches in spread. I have the sticks up and the slack out of the trigger on Janet's 25-06. I squeeze off the round and am rewarded with the sound of a solid hit. Nito seems worried that I have not hit the buck well. He wants me to shoot again. I don't see the need. I pantomime that I'm having trouble seeing the buck through the thick ocatillo cactus brush.

Actually I can see the buck fairly well. He begins to sway, then rears up and topples over backwards. Down and out. It is just 11:30 am. Through my binoculars I can see a lot of antler sticking up. There will not be a problem with "ground shrinkage" on this buck.

I finally achieved my goal of taking a 30+ inch mule deer in the mountains of Sonora, Mexico.

Nito, Jesus and I make our way over to the deer. He is "graveyard dead." A tremendously fitting end to my decades long quest for a 30" mule deer. We load him aboard the truck and head in. Ten minutes into our journey, Nito is indicating to Jesus to stop the truck, excitedly mentioning *los cochis*! Great that we saw some, but these particular piggies do not wait around. They take off for parts unknown. We continue heading in to ranch headquarters.

Santi and Nito head out after lunch with one of the Texas hunters. Doug and I haul my deer to a nearby cactus patch for photos. Afterwards I cape my buck and spend some quality time with Doug. I get the cape all fleshed, handled and salted. Back in the house for a shower and supper. Now I have the challenge of NOT shooting another deer. I can ill afford another $3000 trophy fee, although Doug has "left over" tags that he has purchased for this ranch. One complicating factor: Coues deer tags are expensive too. But they are not near as pricey as a tag for a mule deer!

Cowboy Mexico

Supper is super! Carne asada. Beef marinaded in a fiery salsa and grilled over local mesquite. The Texas hunters are back. They had no shooting opportunities today. One of them is leaving tomorrow. After supper Doug, Janet and Luis retire to the office to discuss the personnel situation. For a variety of reasons, Nito the Skunk Boy is going to be the *guia* who will soon be unemployed. Drawing straws for the drivers, the fickle finger of fate ends up pointing at Jesus.

Pete' del Norte's Spanish Phrase of the day: Que tal estan el arroz y firjoles esta noche? (How are the rice and beans tonight?) Pero commi un chili Serrano la semana pasada. (I ate a Serrano pepper last week.) No perididio nada de su potencia al pasar por mi organismo! (It lost none of its potency passing through my body!)

Dia Numero Tres dawns clear and 44 degrees. Our elevation here at headquarters is about 2,000 feet, which is high enough to be significantly cooler than the valley floor. There is no central heating in the house. Just the huge wood cook stove in the cocina.

With Nito the Skunk Boy long gone, Santi will be my guide. I have decided to splurge for a permit and to seek a good Coues deer. I took a rather nice Coues buck in Hildago County, New Mexico a few years ago. As a cost saving measure, I am determined NOT to shoot one here, unless it is bigger than my New Mexico buck. We spend the morning looking over several deer. None make my self-imposed standards. I'm good with that.... I'm still hunting!

We are back in for lunch and a *siesta*. The remaining Texan talks Luis into saddling a couple of horses and riding out from the house and corrals to hunt locally. I suspect this is a pretty good strategy, since most of the hunting pressure on the ranch so far has been down on the south end, far from ranch headquarters. There simply must be some bucks in closer where hunting pressure has been virtually non-existent. The only flaw I can see with the theory is that since the hunter is a fairly large fellow, they have saddled the stoutest horse in the corral for him. And that horse is a lively five year old stud colt! This is the sort of horse adventure that would end up with me in a hospital. I've seen me do it! They head out.

Cowboy Mexico

Pete Del Norte's Spanish Phrase for the Afternoon: Alto! Dije alto! Por favorcito, alto! (Whoa! I said Whoa! Pretty please.... Whoa!)

After working the capes, Santi and I hunt for about three hours, seeing nothing of note. The Texan survives his afternoon horseback hunt mounted on the young horse. Supper is tasty; breaded and grilled *lomas* (backstrap) from my mule deer. And, for a change, *frijoles*. After dinner activities include some American TV with Spanish subtitles and, Praise Jesus... we are treated to a second Skunk Roundup!

Yet another skunk has appeared on the veranda. Perhaps to avenge the ones that Nito bludgeoned to death a few nights ago. Luis prepares to run the odoriferous beast off. He straps on a headlamp and starts looking around for a firearm. There is consensus among all the assembled hunters that beating skunks with sticks yields less than positive results. Luis wants to kill the skunk from further away than it can spray. But the smallest caliber firearm in camp is the borrowed 25-06 I am using. Additional consensus is that this is at best a "*no bueno*" and might even rise to the classification of "*estupido*!" That is a fairly potent caliber for shooting at 10 feet, especially in close proximity to our living quarters. In addition to the obvious problem of skunk musk flying through the air, there is the potential issue of hot lead bouncing off the *hacienda* walls and patio stones!

Janet has a counseling session with Luis and establishes the ground rules for tonight's Great Skunk Hunt. 1. Not on the veranda. 2. Not IN the house. 3. Not in the shed. 4. Not around Doug and Janet's trailer. 5. Not around the hunting trucks. Etc, etc, etc.... you get the idea. Luis finally figures out that he somehow must most carefully herd *el sorillo* out across the barn yard and parking area and over closer to the corrals. This should be quite entertaining if we can observe the operation from afar.

Luis enlists the help of the assembled Mexican cowboys and guides. He heads out the door with his headlamp and the 25-06. Janet tells me she is "pretty sure he understands the rules...!" But after a minute or two of thinking about it, she changes her mind and goes and retrieves the rifle from Luis. If the skunk steering exercise turns out to be successful, and the skunk somehow is herded out of the immediate area, Janet will consider handing the rifle back over to Luis for application of the "coup de grace."

Surprisingly, no one gets sprayed. Not surprisingly, *el sorillo* is not terribly cooperative when it comes to being herded by the cowboys. He really likes it on the veranda and no amount of poking and persuasion convinces him to head across the yard to the pre-approved slaughter grounds. Still skunkless, but at least with no shooting or "air quality" incidents, the boys give up and head to bed around 8:15 pm.

Pete' Del Norte's Spanish Phrase for the evening: "Barman! Un Pepto Bismol solo! (Bar tender! A Pepto Bismol straight up!) "Se lo agradezco mucho. Y ahora un brindis!" (I am very grateful. And now..... a toast!)

Dawn breaks on another fine day. Luis takes the remaining hunter to El Charco in a truck. Perhaps the dew is off the horsey-riding lily? Which leaves me, at least figuratively if not literally, in "hog heaven." I have 40,000 acres of San Francisco, Lariata and El Pozo Hondo all to myself for javelina hunting! Hopefully we can find those rascals and NOT run into an expensive "*burro*" or a nearly as expensive "*cola blanca*." Santi and I head out. Manuel is the official driver now and we also have Juan along. We head south, stopping to speak with Chango. Turns out Chango has been seeing pigs regularly just south of his house.

Santi knows of a huge patch of *cholla* cactus nearby. *Cholla* fruit and blossoms are a favorite food of both mule deer and javelina. We drive over and soon after beginning to walk the perimeter we hear the distinctive huffing sound of *los chochis*. More experienced javelina hunters than I have described the sound as much like you would expect to hear from a group of gorillas! Santi begins huffing and woofing back at them; he is very good at imitating their vocalizations. Through the cactus I see a big javelina. Santi moves aside and I squeeze off a shot. The pig runs; others follow it. I swing ahead and find an opening. When another pig is square in it, I take another shot. I hear this shot hit; very solid. We make our way over and find both javelina, stone dead.

Santi gets ready for us to pack the two javelina back to the truck. I remind him there are "*dos jovenes*" (two young guys) sitting in the truck. Santi radios

Javelina or los cochis, are members of the wild pig family of the American southwest and Mexico.

the truck and the young fellows show up and pack out the pigs! Santi and I stroll up out of the cholla patch. At the truck we pose the piggies and get ready for the hero photos. But even this becomes a tad complicated. I sit where Santi indicates he wants me to sit for the photo... and promptly get "*nopal*" (prickly pear cactus) spines in my posterior region. All three Mexicans are comfortable enough around me now that they feel they can laugh at me! And laugh they do. Santi even grabs a camera and gets a photo of Juan pulling spines out of my ass with a Leatherman tool!

We haul my pigs back to ranch headquarters. We skin and butcher the javelina and then nap for a bit. Around 4 pm we head out for a last ditch effort to see a buck or two. With limited daylight, we won't get too far from the house. It's a concept I have wanted to try and discuss with Santi, because the area close to the house is normally driven through in the dark, either early morning while heading out, or in the darkness of early evening on the way back in. There just might be a "*muy grande*" hiding in close. Discussing this concept is a little complicated; the language barrier is daunting. Thus I have not broached the subject until now.

Cowboy Mexico

As we drive out, I am paying rapt attention to the hillsides around us. We are only on the far side of the hill with the cross on it, less than 400 yards from the house, when my theory is proven in spades. A mere 100 yards up the hill from the truck stands a positively huge Coues whitetail buck. It is so close I likely won't have time to use the binoculars and then switch over to the rifle scope. I grab the rifle, chamber a round and bring up the scope. The buck is still huge; it was not just a good first impression in the heat of hot pursuit! It has good mass and lots of spread. Plenty of points. I kneel and get a solid point of aim.

The buck knows we are here of course, but he has not moved. He apparently is under the mistaken idea that we cannot see him in the *ocatilla* thicket. The cross hairs are steady just below the point of his right shoulder. For some reason I am remembering that Doug told me that the rifle shoots about three inches high at 100 yards.

I query Santi, just to be sure. *Grande?" "Si, Pete'. Muy, muy grande."* Whispers Santi. That is all the encouragement I need. I squeeze off the shot and the buck drops in his tracks. I bring the scope to bear again, but there is no movement. No need for a "wiggle shot." The truck erupts in whooping and hollering in both English and Spanish.

I have not seen these cowboys this animated over anything since I have been here. The buck is likely as good as I guessed him to be. They have looked at a ton more Coues deer than I have. They would know.

I clear the rifle and compose myself. It's a tall order. I am shaking from excitement and the adrenaline rush. The cowboys race up the hill and yell some more when they get up to the deer. They motion me up and I am subjected to more back-slapping and congratulations.

Back at the house, Janet heard the shot. She calls on the radio to inquire as to what the shot was. I tell her that I have my Coues buck, describing the situation as having shot a "large hole in my bank account!"
After a ton of photos, we get the deer into the truck and head in for a celebratory dinner of grilled pork chops. I head to my bunk at 8:30 pm, chase a spider off my pillow and fall into a deep sleep.

Cowboy Mexico

The Coues is a small subspecies of the whitetailed deer. They are found in northern Mexico and portions of New Mexico and Arizona.

Pete' del Norte's Spanish Phrase of the Day: "No me molestan tanto las arenas." (I don't mind the spiders much.) "Pero prefercia una habitacion sin alacranes." (But I would really prefer a room without scorpions...!)

Since my wallet will not withstand any more successful hunting of my own, the following day I ride along with Doug and Janet in the high seat while Janet looks for a nice muley. We never see a big buck and the major excitement is a braking issue on the truck.

Pete del Norte's Spanish Phrase of the Day: "Cuano se fueron los frenos en este fragment de caca? (When did the brakes first go out on this piece of shit?") "En la epoca de Carranza?" (In the Eisenhower years.....?")

The following day we motor back to the airport in Hermosillo. I say good bye to my cowboy buddies and head home. Most American tourists visiting Mexico head to the beach resorts on the coast. I am much happier inland, in the hills of ranch country, in Cowboy Mexico.

Chapter 12

The Aussies Try to Blow Me Up

By 2007, I had swapped several hunts with Australian hunters Dan Field, G. Ross Ferguson and Jeff Garrand. In fact, as is often the case in these sorts of things, no longer was anyone really keeping track of who owed "what to whom." When we came up with a hunting adventure, we just notified the others, got together and did the hunt. And so it was that I planned to spend several weeks along the gorgeous, picturesque South Coast of Australia, hunting in both New South Wales (NSW) and Victoria. I wanted to see more of Australia and possibly get to chase fallow, rusa or sambar deer.

On Sunday, April 29, I fly into Sydney from Auckland, New Zealand (NZ) where I have just finished up an enjoyable hunt for red stag and sika deer along the Rangitikei River on the North Island. Dan picks me up and hauls me to June Wood's place in Gymea where we spend the night.

The following day we head south along the coast. Past Stanwell Point, and on through Scarborough, Bulli, Dapto, Gerringong, Bonnaderry and Nowra. We have lunch at "Family Fats!" On south through Ulladulla Meroo National Park, Mummerang Forest, Batemans Bay, Mogo, Moruya then out to Narooma. We drive past Mel Gibson's place, but are not able to spot Mel working in the yard or anything!

At long last we pull up to Dan and his wife Liz's lovely self-built house on Corunna Lake and retire to the veranda to rest up before "tea." The sun sets; we can hear the breakers beating on the shore at Mystery Bay. Mullet are jumping in the little bight by the boat dock. Wallabies grazing in the yard. Kookaburras sitting on the rain gage and wattle birds calling as dusk falls. I love this place.

The Aussies Try to Blow Me Up

Dan Field stocking up on firewood in his wood shed. He has a nice wood stove for crisp cool winter nights... in July!

After a couple of days in the "Greater Narooma Metropolitan Area" Dan and I head out to hunt fallow deer on some Forestry NSW ground in a remote area out of Bemboka, NSW. The further we proceed into the Forestry area, the crappier the roads become. There are basically no fresh vehicle tracks from the last couple of weeks. About 11 am, we turn on to a forestry skid road and park. We walk a few yards and thoroughly glass a harvest unit for deer. There are, of course, no deer. We return to the truck and jump in. Dan monkeys with the key and announces "She won't start, mate." Initially I don't worry, since Dan and I are constantly pranking each other and a non-starting vehicle this far back in this block would be a serious problem. It soon becomes evident that Dan is NOT joking. We have a serious problem and are a long ways from any assistance. We have of course, no way to communicate our plight to the outside world.

We check as much of the electrical system as we can. Both batteries, main and backup, are fully charged. The timing chain light comes on when he

The Aussies Try to Blow Me Up

turns the key; perhaps the chain is broken? Neither of us are mechanics or electricians; both of us are plumb out of terrific ideas. Liz doesn't expect us back for four days. Judging from the tracks, no one has been in this harvest unit for at least a week. There is no reason to believe anyone will be by for another week.

Walking out is really the only remaining option. I tell Dan I will be glad to walk with him, but with the nutty Australian gun laws, there are legal implications to leaving a firearm, even in a locked case in a locked vehicle. A plan evolves. Dan, carrying the bolt to the rifle, will start walking out to find help. I will remain with the dead truck and inoperative rifle in its locked case, with no bolt. I suppose in theory this will ensure that errant mentally deranged wombats will not be able to steal the non-working rifle and commit terroristic acts with it. We are guessing that it is about 15-20 kilometers (12 miles) out to the paved road where another vehicle might pass by and give Dan a chance to bum a ride into Nimmitabel. Here he might find a phone and call the Aussie equivalent of AAA, the NRMA (National Roads and Motorists Association. Insurance with road service.) I get comfortable and remind Dan to be sure and take his heart medicine with him!

It's about noon when Dan strides purposefully off down what passes for a road here in the State Forest, toward civilization. He has his day pack, rifle bolt, a small first aid kit, some snacks, two liters of water and his wallet. I settle in for what will likely be a long wait. My thoughts turn to wondering how in the world Dan will make this journey without any beer.

It's possible Dan could be back with help tonight, but I seriously doubt it. I would actually be surprised if his tired old ass makes it to the pavement by sundown. It would be a real shocker if a vehicle were to pass by my location. But on that off chance, I move a few things from the truck out the 100 yards to the logging haul road. At first I take only some snacks, a coke, a deck chair and a paperback book. A bit later I erect a tripod of sticks and tie some flagging onto it. I decide that come about 4:30 pm I will put up one of the tents and gather a bit of wood for a fire for the night. I have all the groceries for the trip; Dan has a few snacks. I have a soft chair; Dan has a 15 km walk. I'm pretty sure I have the better deal.

The Aussies Try to Blow Me Up

Lo and behold, just before 3 pm I hear a vehicle approaching. Perhaps it is just wishful thinking, or a mirage…. In truth, it is an NRMA tow truck with Dan riding shotgun sporting a big grin on his face. The truck pulls up and stops. Its operator is NRMA mechanic, Ross. "G-day Mate! Glad to see you! 'How ya goin'?"

Dan explained to Ross that this is a serious emergency. He told Ross he has been forced to leave a ditzy, nervous American tourist "out in the Bush" where he might well be devoured by dingos or ravished by rabid wombats. Thus it was imperative to rush back to the site of the breakdown. With the story as told by Dan, implying that the tourist could go off the deep end at any given moment due to being alone in the wilderness of the New South Wales bush, Ross has been rushing to get here. As Ross rounded the corner and pulled up, a somewhat different scenario played out.

I am kicked back in my comfy camp chair, shirt off, soaking up the sun, puffing on my pipe, feet propped on a full cooler, with snacks and a cold Coke close at hand. I'm reading my paperback book. Ross remarks to Dan: "To me, your tourist doesn't really look real close to death or even overly concerned!" Dan tells him that looks are deceiving and "the bloke could 'snap' at any moment!"

Dan's legendary good luck has served him in good stead once again. He only walked a few kilometers before running into a Forestry NSW "dogger" (Feral Animal Control Officer.) The dogger had a sat phone in his ute and Dan was able to call NRMA. Not only that, but Ross the NRMA Mechanic just happened to be up on the escarpment above Bega, within 10 kilometers of us, looking for some stranded tourists who had called for rescue themselves. Ross drove right over, picked up Dan, and came straight on in to the scene of our breakdown. This sort of luck is obviously a testament to Dan's record of clean living. What else could it be?

Ross works his magic on Dan's SUV. As it turns out, it is not a timing chain, just a stuck relay in the electric system. A few taps (in precisely the proper spot…) with a spanner and we are running again! Ross and his tow truck disappear in a cloud of dust; we are back on our way again. Not only are

The Aussies Try to Blow Me Up

we mobile again, we are armed with some good intel on the deer situation, courtesy of the "dogger" who rescued Dan. For starters we go back and locate a "log dump" (log landing and sort yard) back a couple of kilometers. The "dogger" had pointed this out to Dan as a good potential camping spot with plenty of firewood. From there we proceed out the haul road a ways to a major spur. We drive a short distance and park the truck. With more than a little bit of trepidation, Dan turns the truck off (and yes, it restarts just fine…) We walk the rest of the way into Packer's Swamp where the "dogger" says there are a lot of fallow deer.

We hunt the remainder of the afternoon and much of the next day without seeing much deer sign. Nor do we see any deer. To minimize our losses we drive back out to Nimmitabel and call to warn Liz that we are coming home. Finally, we tell her we only had to be rescued one time!

After spending a few more days enjoying the sights of Narooma, I play tourist in some of the shops and spend a morning talking to Dan's daughter Zoe's elementary school class about Alaska. We fish a bit and do some crabbing in Corrunna Lake. I have a nice discussion about deer hunting with Zoe. Zoe was only a month or two old when I last visited. She has now successfully shot a fallow deer herself. She asks me if I will be shooting a stag bigger than hers? I tell her that I hope so. She proclaims "That's not what my Dad said!" By now Dan has come up with another hunting adventure. What could possibly go wrong?

We head south to the Eden area, stopping to visit a bit with Clyde, a hunting mate of Dan's. Then on to the farm of another mate, Kurt. Kurt has a bunch of fallow deer on his place, and a few rusa as well. His property is fenced, but all the gates on the State Forest side are open. Deer come in from the public ground because Kurt has better "groceries" for them. Over the next couple days we are able to kill three nice fallow stags. I do indeed somehow manage to kill a stag bigger than that of a sixth grader! With coolers full, we motor back to Narooma.

Dan has some work to do, so he is pawning me off on Jeff Garrand and his boss at work, Steve. Steve is otherwise known as "Staino." It's a reference to his stainless steel fabrication shop. Jeff, Staino and I will be heading down

The Aussies Try to Blow Me Up

I spent some time talking with Zoe Field's sixth grade school class. What great questions!

into Gippsland, in northern Victoria, to hunt sambar deer from Steve's mountain cabin.

Soon we find ourselves near Buchan, Victoria, in northeast Gippsland in the Snowy River drainage. In fact we travel through some of the residual bush (wildland) fires that have been ravaging the area since February. We make it to camp by 10:30 pm. It's pretty chilly here in the mountains, but we warm both cabins with toasty fires in the wood stoves. This is a pretty classy hunting cabin. It has a gravity fed water system, on-demand propane fired hot water supply, a flush toilet and a shower of sorts. By midnight we are in our bunks.

We spend the next morning making a good recon and scouting expedition. We walk some thickly vegetated gullies choked with blackberry vines. The wind is terrible and we see no sambar, but from the looks of all the sign, they are here. We quit at mid-morning; the sambar are bedded for the day. Next we tackle

The Aussies Try to Blow Me Up

a firewood gathering project. We cut some dead hardwood, mostly gums, and haul the wood in Steve's Kubota 4WD, side by side ATV. Back at the cabin we split and stack our bounty. Around 1 pm we partake of a chunk of lamb that has been roasting in a Dutch oven since dawn. After lunch I head out to sit for a deer. Steve and Jeff work on fashioning a drip-torch of sorts out of a bottle and some fuel line. They hope to burn some brush. It would appear the rules for private use of prescribed fire are a tad looser here than back home in the states! Their attempts result in no fire of any magnitude. I postulate the humidity is too high for decent chances of ignition.

For the evening hunt, Steve drops me at the top of a thick gully. He will attempt to push sambar to me. His last advice is to not shoot "the house cow (sambar hind.)" I don't. In fact I never see a deer. I amble back in to the cabin at suppertime.

Steve has been trying to grow some citrus trees here on the place. When one of the first saplings died after transplanting, he went back to the nursery for a replacement. This time, instead of a larger version, with a root ball, the nursery gives him a smaller one. This is what we foresters call "bare root stock." Steve is skeptical. He asks about how to treat this new tree. The sales lady says to apply urea to the soil. Steve seeks clarification: "So basically you have sold me a 'stick,' and told me to piss on it...?" Yes, that is the essence of the instructions!

We have some quick "tea" (supper) and head out to procure a bit of dogfood (in this case, kangaroo meat) for Nessie the dog. The weapon determined to be best for the task is Jeff's Brno in .308 with hand-loads he recently worked up, loaded up and test fired. Jeff stays in camp. Steve and I proceed out toward the gate. Our first kangaroo is quickly spotted. I bring the .308 up to my shoulder, center the crosshairs, squeeze the trigger and... suddenly my world changes.

It is instantly obvious something is radically wrong. Instead of just muzzle flash, there is a terrible explosion and I am simultaneously in tremendous pain. I cannot see anything. Blood is running down my face and dripping onto my lap, boots and the ground. At first, neither Steve nor I have any idea

The Aussies Try to Blow Me Up

This is the .308 rifle that exploded and began my ordeal that ended in eye surgery.

of what has happened. It appears the rifle has "exploded." A big chunk of the left side of the wooden stock is gone. The floor plate is bent into a U-shape and the shells from the magazine have fallen out. The bolt is "frozen" in the closed position in what's left of the action.

My face is numb. My eyes are burning and tightly shut. I cannot see anything. Steve observes that my face is a mass of small cuts, many bleeding profusely. I grab my handkerchief and out of reflex, jam it against my face and eyes. In retrospect, this was an extremely poor idea. It seems there was a large wood splinter sticking out of my cheek just below my right eye. Neither of us was aware of its existence. When I slapped the handkerchief up, it jammed the sliver of wood through my cheek. Later we find out that it punctured the sinus cavity below my right eye. I drop the handkerchief, feel for the sliver, pull it out, and reapply the cloth. I'm still essentially blind. Steve wheels the buggy around and heads for the house.

The Aussies Try to Blow Me Up

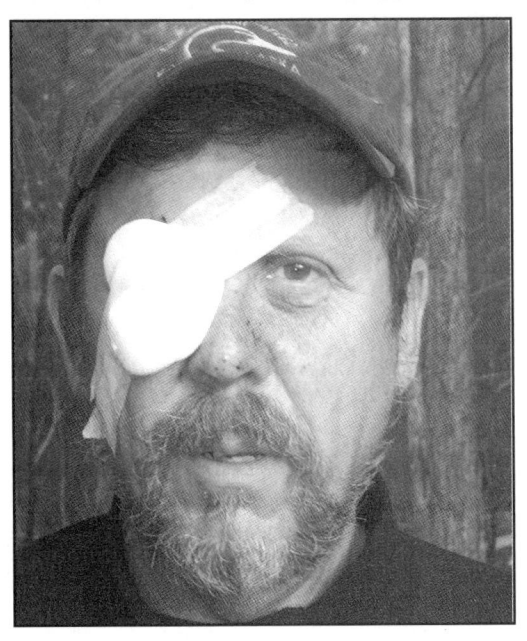

My Aussie EMT friends took good care of me, flushing out my eyes and patching me up for transport to a series of clinics and hospitals.

I am lucky in one sense. Both Steve and Jeff are firefighters and the Aussie equivalent of EMTs with the NSW Bush Fire Service and the Narooma Rescue Squad. As first responders, they are trained in all sorts of first aid. They have a really skookum professional first aid kit. But about the best they can do is flush out my eyes and apply some ointment to the lacerations in my face and hands. Clearly my eyes have a lot of burned gun powder embedded in them. That much is obvious. Later we come to learn that I also have wood splinters and shards of brass shrapnel embedded in the corneas in both eyes. The right eye is in worse shape than the left; this was the side closest to the action on the rifle since I am a right-handed shooter.

The nearest medical facility of any kind is a rural clinic in Bairnsdale. It is a long way away over some terrible roads, but we need to go. We load up into the truck and head out to the paved roads and on to Bairnsdale. We are at the clinic shortly after 9 pm. At this time of day there is no doctor present. I am seen first by the triage nurse, and later by two interns. There is little

The Aussies Try to Blow Me Up

they can do aside from additional flushing of my eyes. They turn me loose with some eye ointment and pain pills in hand. I am grateful and pay with my Visa card; my American medical insurance does not work in Australia! The interns encourage me to return in the morning when the doctor comes in or, better yet, head for the hospital at Bega, north back toward Narooma. We drive the bumpy obnoxious roads back to camp. By now I'm in even worse pain.

The next morning we reassess the situation in the light of day. Among other revelations, I find I have quite a few bits of burned gun powder and walnut splinters in my right hand. More are showing up on my face. Inflamed and swollen, this is a face that even a mother would have trouble loving! We discuss returning to the Bairnsdale Clinic, but decide instead to head back toward Narooma. Jeff and Steve button up the camp and we leave around 10 am. We stop for a tailgate lunch and "cuppa" around 1 pm. Steve pulls into the Bega Hospital at 3 pm. We also call Dan and Liz to warn them we are returning. The Bega Hospital has a young doc on duty in the ER. Bless his heart, this is his first day on the job. He is overwhelmed by the idea of being presented with an eye injury. He recommends that I see an opthalmologist. I agree.

From Bega, we drive back to Narooma. We are at Dan and Liz's home by 8:15 pm. Liz helps me get ointment into my eyes and I hit the hay after emailing to let family in Fairbanks know about the injury. The next day Dan is going to haul me to the ER at the big hospital in Canberra, ACT (Australian Capital Territory). The ACT is much like our Washington, DC. The seat of national government, yet not actually its own state. My eye is still hurting a lot, plus, now it's extremely sensitive to light. After check-in, we wait only 20 minutes before being called for the triage interview and moved almost immediately into a specialized eye-trauma treatment room. The young doctor is sympathetic, compassionate, professional and smart. He immediately opts to ask for the services of the on-call eye specialist.

When he leaves to do this, I am treated to a regular floor show; Dan just cannot help himself. He gets to messing with some of the eye exam equipment and inventorying the tools and dressings. He even sets himself up behind the Topcon SL-3E Slit lamp microscope and has a look at the bits of burned gunpowder embedded in my right thumb. Interestingly, some of

The Aussies Try to Blow Me Up

When I hear the term "bull in a China shop," I often recall Dan fooling around in the eye exam room at the hospital.

his comments make sense. He claims he can see not only powder fragments, but also some brass shards and fibers of some sort. All embedded in the skin and flesh of the web between my thumb and palm.

On the wall are two eye charts. One is the usual one with letters on it, featuring a big "G" at the top center. The other has common (for Australia…) animals and objects on it. It is apparently for either kids or really dumb people who do not know their alphabet. Dan has fun with it; wonder which category he fits into? Having Dan with me, unsupervised, in a doctor's office gives me flashbacks to the old "bull in a china shop" analogy!

Dr. Alex comes in and has a good look while we wait for the ophthalmic surgeon. The more he studies my eye injury, the more Dr. Alex is glad he called for reinforcements. He eventually goes off to see some other patients. I stay seated in the exam chair. Dan collapses on the gurney and has a nap. Luckily I have my camera with me!

The Aussies Try to Blow Me Up

The doctor shows up and spends a good half hour examining my wounds and injuries. I must admit it is somewhat unnerving when he first peers into the scope and exclaims: "Jesus Christ....!" He reports that he is seeing "multiple" bits of burned gunpowder and at least two large shards of brass embedded in my right eye. One piece of brass is quite deep. The doctor hopes that surgery won't be necessary, because moving that shard might cause it to puncture the eyeball and cause a leak of fluid. He wants to see me in his clinic, here in Canberra, on Monday.

Dan asks if he can take a photo of the two nice doctors and me. "It's for the lawsuit" he says! The younger doc turns a bit pale.

We go back to Narooma for a day. The following morning Jeff hauls me back to Canberra. Now the doctor wants to refer me to another, higher level, facility. And thus, two days later I am checking in to the Royal Victorian Eye Hospital in Melbourne. By this time I have consulted with my regular eye doc in Fairbanks, Dr. Ruth, as I am fond of calling her!" Ruth has looked over the report from Dr. Dixon and has been able to give me some advice via email from afar. I'm becoming more and more convinced that some sort of surgery is in my immediate future. Both the surgeon and Dr. Ruth have advised me to seek help in Melbourne if the pain persists. And "boy howdy," it is definitely still very painful!

Ross Ferguson has driven me to Melbourne. He is with me when a team of corneal specialists bluntly tell me that surgery will be necessary. I discuss it with them. They are only giving me 20 minutes to make up my mind as to whether to have the surgery done here. Things are moving way too fast. After they assure me that air travel and pressure changes "probably" won't make things worse, I opt to go ahead and return to the States and have surgery there. Within four days, I am winging my way from Sydney to Fairbanks.

Once home, I see Dr. Ruth, but within a day or two I fly from Fairbanks to Anchorage and undergo the needed surgery on my right eye. While I am "out," the docs remove the shards of brass and the wood splinters. Fragments of burned gun powder, it turns out, are something that can safely be left embedded in my cornea. Who would have guessed? The incisions are sewed

The Aussies Try to Blow Me Up

My nice Spanish fallow stag taken in southern New South Wales.

up and I fly home. The eye bothers me for the remainder of the summer, but in early fall, when the sutures are removed, I am relieved to find I have regained nearly 20/20 vision. I am one very lucky guy.

To this day no one is sure what actually caused the rifle to explode. The most common theory is to blame the hand-loads. Perhaps it was a "squib load" with very little powder in the case. Gunsmiths who examined the action and the shell casing think perhaps when I pulled the trigger, the "fire" from the primer burned right over the small amount of powder without immediately igniting it. This may have caused the bullet to travel partially up the barrel and stop. When the rest of the load of powder finally ignited, it was into a blocked muzzle. While the bullet did exit the muzzle eventually, a lot of pressure was built up in the action, causing the action to explode.

My Aussie mates spent a lot of time worrying about my injury and subsequent surgery. I am grateful to them for taking care of me and running me around

The Aussies Try to Blow Me Up

Jeff and Staino working up some terrific tucker in the Dutch oven at camp in the Stony River country of Gippsland.

for medical care. I, of course, spent an inordinate amount of time reassuring them it was not their fault, and I had made a complete recovery. So when I returned to Australia a couple of years later, I decided to have some fun at their expense. I procured a telescoping red and white "blind man's cane" and hauled it in my carry-on. As I deplaned, I had on huge sunglasses and was tapping ahead of my route with the cane, doing my very best Stevie Wonder impression. If I could have borrowed a German shepherd dog to imitate a Seeing Eye dog, to complete the spoof, I would have. Dan, Ross and Jeff about crapped their pants when they saw me exit the plane and "feel" my way to the greeting area. It was GREAT! Somehow I imagine there will be payback though! In spades no doubt!

Chapter 13

My Name is Tumkulu!

After hunting a time or two in northern South Africa, I thought it would be fun to try expanding my horizons and see some other areas. African hunting is indeed an addiction, and South Africa itself is a large country with many different types of habitats. The Limpopo Province in the north is interesting and game-rich, but fairly flat and non-descript. The brochures and websites describing the Eastern Cape region of the southern part of the country showed not just rolling, but mountainous country. Additionally, it had a number of species that I had not hunted in the northern part. In late April 2013 I headed out (in a blizzard) from Fairbanks, bound ultimately for Port Elizabeth, a lovely seaport on the Indian Ocean. Rather than do this new adventure on my own, I bamboozled several buddies into coming with me. Old stand-by mates Dan Field of Narooma, NSW, and G. Ross Ferguson of Wangarata, Victoria were representing Australia in the group. And from the State of New York, Frank "Big Hat" Rizzuto and his 13 year old son, Hunter. We would have the place to ourselves. This was good, because anyone overly sensitive, easily offended or without at least a three-star sense of humor would have been uncomfortable around this bunch.

As is my habit, I am flying on Alaska Airlines using "miles" instead of money. For the miles I have expended, I am flying First Class on Alaska Airlines to Seattle. Then First Class on British Airways (BA) from Seattle to London to Johannesburg. First Class on Alaska Airlines is one thing. On BA, "First" is a whole 'nother level of wonderfulness. Seats are luxurious, and set in "private" cubicles, a fitting tribute to a tall traveler. I'm in such luxury I can barely remember my first flight to Africa in 2003, in coach with Peyton, on South African Airways, ten years prior. It was somewhat uncomfortable, best as I can vaguely recall. Flying in Coach is an ordeal for people of my tall, overweight stature. Flying in First Class makes the long international flights an integral part of a wonderful adventure.

My Name is Tumkulu!

Modern safaris in Africa normally include daily laundry service. Most of the time you get your own clothing back!

I sleep most of the way to London. There I deplane, get through more Security and move to Connecting Flights. One more trip through Security. Out the other side, my sagging trousers are what, in the business we call a "clue." A clue that I left my belt in a plastic tub on the table at Security! Bad enough that I am suffering droopy drawers. The belt is a money belt which contains cash to pay the balance of my safari cost. Also in that tub is a pair of $8,000 hearing aids. Security may have been paging me, but for some reason I didn't hear them. Once reunited with the remainder of my belonging I head up to the British Airways "Concorde Lounge." First Class on BA will bring a tear to your eye. You have access to the Concorde Lounge which means seated dining, at no additional charge. The attendant has looked at my boarding pass. She notifies me when it is time to board my flight to Joburg. I stroll to the gate and onto the aircraft. We are wheels up at 6:35 pm. Of the 14 available cubicles in First Class, only five are occupied. Four high rollers and.... well, little ol' me!

My Name is Tumkulu!

I have not had a ton of luck sleeping on this leg. At 6 am the cabin lights come on and breakfast rolls out. According to the GPS map on my private video screen, we are somewhere between Lusaka, Zambia and Windhoek, Nambia. I open the window cover and am greeted by an incredibly gorgeous red-tinged sunrise poking over the eastern horizon. Reminds me of Hemingway's reference to "MMBA." Miles and miles of bloody Africa.

We land at Johannesburg. I make my way through Passport Control without incident. My checked duffle bag shows after just five minutes of waiting. I head out into the greeting area where old friends Marius and Anna rescue me and whisk me away. Within a half hour I am comfortably ensconced at Afton House and having another breakfast. There are several other Alaskans in the house this morning. The most fascinating guest is a hunter from Arizona who was with PH Erwin Kotze when he was killed by an elephant last week on the Caprivi Strip in Namibia. I read about the incident, but hearing it from someone who actually witnessed it (and seeing his photos…) is both vivid and chilling. Most of these hunters move along to their ultimate destinations. I spend the day sunning and resting up.

The next day I'm up early to accompany Marius to the airport to collect Frank and Hunter. They arrive on time and get through the South African Police (SAP) gun permit process. Back to Afton House to await the arrival of Ross and Dan. I return to Tambo International Airport at mid-afternoon with Marius. Four Americans bound for Afton House come out first. Marius takes them to the SAP office. I wait for the Aussies. I have a perfectly good printed arrival-pickup sign from the Afton House office with their names on it. But I flip the sign over and craft some new wording more to my liking: "Welcome Old Ross and Fat Dan." There, that should work much better. It DOES get their attention! So wonderful to see my mates and be the first to get a shot in!

We have a quick reunion, slide by the SAP office to clear the weapons permits and change some Aussie money to US dollars and RSA Rand. Marius hauls the Americans to Afton house. When he returns he has Frank and Hunter in tow. He drops the lot of us at a nice Portuguese fish place called Jose's. As

My Name is Tumkulu!

always, dinner with this bunch is part of the adventure. The waiter drops a bottle of Castle Lager on the edge of the table. It breaks and spills beer all over my pant leg. Now I smell a lot like Dan... We have a terrific dinner of fish, prawns, mussels and calamari. A few beers and a few Cokes.

We arise early after a night of happy anticipation, and repack for today's domestic flight to Port Elizabeth (PE). Marius hauls us all to the airport around 6:30 am. There are issues with SAA regarding Dan and Ross's ammo. Although being in locked metal cans, it must be extracted from their duffle bags. It needs to be checked separately, with an extra bag charge of course! Going through the Security line and x-ray is always challenging and today is no exception. But we make it and find seats at our gate for the flight to PE.

Entertainment is provided by a young lady with about eight progeny in tow. They are winding their way through the railings leading to the waiting area. One child, a boy of about six, gets busy inspecting something he finds interesting and falls quite a ways behind the rest of the brood. He races to catch up and tries to cut a corner by ducking under a metal rail. Much to his chagrin, he finds the space between the floor and the railing is actually blocked by a very clean piece of plexiglass! With a loud crash, he impacts it with his forehead and bounces off. We try hard not to laugh; it's difficult to remain composed. He is not so lucky with his siblings. They laugh uproariously and loudly mock him all the way to the door of the aircraft.

Into PE on schedule. Outfitter Gary Sparrow and PH John Barnes are there to help us collect and assemble our luggage and firearms. We pile into their two bakkies and head out toward Grahamstown. Once there we fill with petrol (gas up) and pick up a few groceries and sundries. At the end of the pavement is plenty of gravel road. We finally turn onto the farm road into Waterfall Farm. This is a picturesque cattle and sheep family establishment in delightful hilly country. Frank, Hunter and I will be in the bunkhouse. Dan and Ross will occupy the fancy quarters in the new lodge. We get to meet the ranch owners, Kevin and Natalie, a number of staff and quite a collection of canine companions.

My Name is Tumkulu!

After a great supper, we stagger off to bed. My nocturnal bliss is interrupted by Frank plugging in too many appliances in his room, thus knocking out the power to the entire building. John comes down and flips the breaker. We have power again.

Day One of our safari is clear and 44 degrees. A bit more brisk than what I expected. What North American would have guessed that in the southern hemisphere, the further SOUTH you go, the cooler it is? Clever fellow that I am, I have brought some light long johns with me. The lack of any central heat in the sleeping quarters is an interesting issue. It's not critical, nor do I need heat at night. But it sure would be nice to have a bit of heat when going to bed and then again whilst getting up in the morning. I have a great night's sleep nonetheless. We get up, have breakfast and head out to meet the trackers and sight in our rifles. The Ruger .300 WinMag John is furnishing me is a very nice firearm. John will be my PH. I am happy to see his tracker is a "grey beard." Nahoy is an ancient man of the bush! He has hunted with John for 10 years. They have a close, friendly and mutually respectful relationship. On the negative side, I won't be BS-ing with Nahoy very much. He speaks his native Xhosa and some Afrikaans. English is fairly foreign to him.

Once the rifles are all OK, we all head out to a neighboring farm in the conservancy to hunt black wildebeest. We hunt the open high elevations. Everyone except Dan manages to take nice wildebeest trophies. We are back at Kevin's in time for lunch. In the afternoon John and I pick up Nahoy at the skinning shed and head out to a different, rougher area of Waterfall farm to look for mountain reedbuck. We spot several reedbuck, but the only trophy ram fouls up our stalk and disappears in the brush. The most notable sighting of the afternoon is that of an aardwolf. It's a fascinating insectivore I've not seen before in my travels in Africa. Sort of like a long-legged South African wolverine.

We are the first ones back in camp. I go to my room. Stacked on my bed is all Frank's clean laundry. I take it over to his room. Sure enough, my stuff is neatly stacked on Frank's bed. Daily laundry service is standard fare in African hunting camps. But accurate sorting sometimes escapes the staff's priority list.

My Name is Tumkulu!

Madala Barnes with his trusty tracking dog Spud.

Once the laundry roulette is thrashed out, I go up to the office and partake of the wifi. How spiffy is that? A safari with Wi-Fi! John Skypes with his wife Joan in St. Francis Bay. Spud communicates in some form with his canine girlfriend, "Minnie Mouse." Ross and Eddie roll in. They saw zebra, but were afforded no shot opportunities. I predict Ross's long run of bad luck on zebra (four safaris in two countries) will soon be over. After dark, the other hunters roll in. Now all four adults have their black wildebeest trophies. Dan's is a magnificent old bull, cleanly taken with the old 1876 Sharps black powder rifle. We partake of a mixed grill over the braai. John urges me to get ready for the next day's adventure. He and Spud and I will be hunting nyala on a different property some distance away.

The next morning is the first of May. It is chilly again. I slip out of the long johns I slept in and into a pair of shorts and a down jacket. I hope that temps will rise enough that I will not be a victim of hypothermia; it is supposed to be 80 degrees by noon. I get over to the new lodge before the others. I catch

My Name is Tumkulu!

My tracker friends Nahoy and Sandile and I with our hard-earned nyala bull.

up on my email and have a bit of breakfast with John and Spud. We hear Dan and Ross arising. While Dan is in the bathroom, I add a bit of salt to his coffee. Dan must be sleepy. He drinks it right down and says nothing. Next time I will add more salt. Probably better that he didn't notice; payback can get dicey when Dan and I begin to compete in the practical joke department.

John and I pick up Nahoy and hit the road in the Rover. It takes an hour and a half to reach the other farm, Woodale, over some fairly crappy roads. I originally understood it to be closer. At one point I ask John: "How far are we going?" He answers the question with a question of his own: "Do you have your passport?"

We enter the yard on Woodale around 8 am. As we pull up to the barns, there are a few workers about, but no sign of anyone higher on the food chain. We pick up our Woodale farm worker, Sandile, and head out. South African law requires when you hunt another's property, you are required to have one of their staff with you. It's a good idea, especially when it comes to knowing the ground and having keys to the gates. John has me

My Name is Tumkulu!

pumped up with tales of the last time he hunted here. "I had seen six nyala bulls before 8:30 am." Yeah? Well that was then and this is now! We are travelling through a nyala-free zone. Just short of one of the boundary gates, a young nyala bull crosses the road in front of the bakkie. We stop, park and head into the virtually impenetrable thorn bush. From a new vantage point we can see giraffe, and then some kudu cows filing up out of the gully. Then another immature nyala bull.

Suddenly we hear something racking brush. Whatever it is, it is fairly close, but completely obscured by vegetation. Below us, some 60 yards away, a magnificent bull nyala appears momentarily in a tiny opening. John can see only spiral horn tips. At first he suspects it is a young kudu. Then, with a start, we both suddenly realize it is a very fine nyala bull. John tries to set up the sticks, but the hillside is steep. Putting up the sticks does not work well at all. "Shoot him" orders John. I try my best to comply. Quickly finding a decent sight picture on what few unobstructed pieces of nyala are showing, I squeeze off a round with the .300 WM. The bull is shot through the spine at the juncture of the neck and thoracic cavity. He doesn't even kick. The clean kill, which is always a good thing, is doubly wonderful here because of the thick vegetation and steep terrain. Following a wounded animal here would be very problematic.

John and I hike back to the bakkie for our cameras. Sandile thinks we can get the bakkie fairly close by crossing the drainage further out on the track where we are parked. From there we should be able to cross a dam and follow a narrow track down toward the kill site. We park when we are as close as we are going to get. Nahoy and Sandile grab "loppers" and a panga and cut a narrow trail from the track to the downed animal. We walk up to the bull and John proclaims that it is a "corker." It looks big to me, but I have exactly no experience with nyala. I'm torn between wondering if it is really a very good trophy or whether John is blowing "PH smoke" on my behalf. Nahoy, Sandile, and John all keep uttering the Xhosa word "nkhulu" (big). As time wears on, I come to the conclusion that it really is a very nice trophy. Regardless of how it stacks up against others, I am extremely happy with it.

My Name is Tumkulu!

Nahoy and I glassing for game. We could spot game together; we just could not verbally communicate!

We haul the bull back to the skinning shed where Nahoy and Sandile handle the caping and butchering chores post haste. The meat goes into the walk-in cooler. As is the law in South Africa, the meat belongs to the landowner. When we head out, we will have just the cape and horns but the meat will not go to waste. I watch and take photos. Spud wanders about the yard, ultimately making the bad decision to run his wet little nose on the "hot" electric wire on the fence surrounding the house. This results in a loud yelp and a somewhat deflated ego on our little "inja" (dog).

We are packed up and ready to leave Woodale soon after 11 am. We are back at Waterfall Farm for lunch by 1 pm. Ross is in camp with a very nice black springbok. As he is wont to do, he has picked up several sundry skulls of springbok, porcupine and even a red hartebeest. After lunch I take a cool shower and lay down for a bit. Soon I am up and trying to find my tools for cleaning my hearing aids. Q: "What do you call a really dirty hearing aid?" A: "An ear plug!"

My Name is Tumkulu!

For the evening hunt, we head out to the far reaches of Waterfall Farm looking for some of the smaller antelope species. John refers to them as "pie fillers!" I get to play with a 25 pound leopard tortoise at the old kraals. By 4 pm I have a fine black springbok added to my collection. We head in early. Nahoy has earned some "down time," and we drop him at his hut. John and I process the springbok.

Clearly John and I are getting along quite well. We have the same droll sense of humor. He seems to trust me and my firearm handling ability (at least I haven't killed anyone yet). He has seen I can shoot. We are becoming firm friends. I feel I can ask him a personal question. "The staff all refer to you as "Madala." What does this word mean?" John explains that it is Xhosa for "wise elder." I'm pretty sure he is telling the truth. He is the oldest person in this operation. And he certainly has earned the willing respect of all the others, both white and black; it seems suitable. I pursue the issue further. I tell him I need a Xhosa nickname myself and ask for his help in choosing something suitable. John consults with Nahoy and our other tracker, Umbutu. With conspiratorial smiles, the consensus is that I will henceforth be known as "Tumkulu." John assures me that this means "tall man." I like it! While working in the skinning shed, I hear occasional mention of my new nickname by the assorted trackers and skinners. I'm no dummy; they are laughing and snickering AT me! Clearly there is skullduggery afoot here. When I get back to the lodge and have access to my computer and a Xhosa/English translation website, the plot is exposed. My new nickname actually means "tall, fat man!" As suspected, I have been royally "had." But it is all in good fun and the name sticks. Too late to select a new nickname. I will bear this one proudly.

The next day is much warmer. I laid on top of the blankets all night. After a quick breakfast, John, Nahoy and I head out on foot from the lodge for another try at reedbuck. It doesn't take long to find a small herd of them. They are within range. I rest the rifle on the sticks and take a poke at the ram.... and miss. We watch the 10 females for nearly 30 minutes before the ram reappears. Turns out he was much closer than I estimated; I shot clean over him. The reedbuck move off, barely spooked. The ram shows again in the clear at 160 yards. Despite my crappy shooting a few minutes ago,

My Name is Tumkulu!

this time I hit the ram squarely and he goes straight down. Done..... with a trophy reedbuck "in the salt," before 8 am.

With plenty of daylight left, we head over to Cloudlands to try for a common springbok. I notice Kevin is hunting with Dan on the top end of this same pasture. The animals here are a tad jumpy from all the hunting pressure. We decide to head out to a more distant part of the property. Back at the bakkie, John gives Spud a big drink of water. Spud consumes it with relish, then burps like a man! Again like a man, he proceeds to fart. As John so aptly puts it: "Spud is running a bit rich today!"

We drive back in for lunch. Ross has broken his 10 year jinx; he has his zebra! Dan suggests that Ross's PH, Ed, will probably be in Ross's will now! Ross has had terrible luck on zebra in the past.

I tag along with Dan and Kevin for the afternoon. Dan is still armed with the old Sharps, so when a good springbok is spotted way out beyond the range of the old gun, I am offered the opportunity to "reach out and touch him" with the .300 WM. I make the 280 yard shot in fine form. Missing a shot while Dan is watching and videoing could have resulted in DAYS of harassment!

Frank and Hunter have made a scene at the pool in front of the lodge. Frank was standing on the edge of the pool and it was just too tempting for Hunter. Momentarily forgetting who picked up the tab for his first African safari, Hunter pushes Frank into the pool. With the cool evening temperature, the idea of a swim is not terribly inviting. John describes it in Afrikaans as being "leeu koud." I make the mistake of asking what this means. Translation and further explanation reveal the weather is so cold, that a very little bit of a man's exterior plumbing shows, with a "mane" around it, thus looking like a lion! I had to ask, didn't I? Ah, local color, in all matters!

The next day finds us heading out to Simon Hawkins' Three Fountains Farm to hunt springbok. The animals here are a lot less spooky than those on Waterfall and Cloudlands. These springbok are so relaxed, even Dan is able to shoot one, as do Hunter and Frank. We pack up and head back over to

My Name is Tumkulu!

Waterfall and out for the evening into a steep, deep valley that drains directly into the Great Fish River. It's a badass two-track road that gets progressively worse as we ease down it. We park the bakkies and leave the trackers there, while we walk on. A few yards down the track we come upon a small tortoise. Dan wants some photos and gets same. Then, he picks up the tortoise. The surprised animal hisses at him. Dan screams like a little girl and tosses the poor tortoise into the air, whereupon it drops to the ground. This startles Spud who is standing right behind Dan and bumps into him. This surprises Dan and again he screams like a little girl (and likely pees down his own leg). If only I could have been ready to video this show. Meanwhile, the laughter of the group has probably spooked any and all game out of this area. We sit and glass until dark, then move back up the hill to the bakkies where we left Nahoy and Umbutu. It is starting to rain. Back at Waterfall, we go to our rooms to get ready for supper. The laundry inventory is messed up again. My favorite khaki hunting shorts are just plain GONE. No one in our group has them.

The next day is a "non-hunting" day per our contracts. I go over to the lodge early anyway, since I intend to Facebook message Emma Lee and Tucker. I wish I had thought to bring in some firewood before it got rained on last night! But I take a knife and shave off some splinters from a stick of wet wood and get a fire going in the fireplace. Dan gets up and staggers to the coffee pot. He is fully as ugly as when he went to bed. The wind picks up and the rain increases. It is supposed to clear around mid-day, so we load up in the bakkies and head toward Kwandwe Game Preserve anyway. It's only a 45 minute drive. We are welcomed by several bundled up but smiley staff and presented with steaming mugs of hot chocolate. A jovial fellow named "Gibson" loads us into a covered touring car and hauls us to the main lodge.

Our little group makes a grand entrance at the Great Fish River Lodge. It is quite spectacular and obviously set up to cater to the "eco-tourism" crowd. We partake of a gourmet lunch with wine, beer and lots of fun fancy stuff! Hunter is really impressed. Dan and I have fried calamari. For dessert, we partake of tiny slices of grilled pineapple with two drops of caramel sauce and a sprig of mint. What itty bitty portions; I'm hungrier at the end of the meal than when I started!

My Name is Tumkulu!

After a bit our "driver/guide" shows up to collect us for our wildlife viewing tour. "Doc" has a gun case with him. He says it's a .375, but it remains in the case and we never see it. The case may hold an "assegai" or even a broomstick for all I know. It certainly would have been difficult to bring to bear if we needed to be saved from rampaging elephants or charging lions… The tour itself is not bad, but not spectacular either. This is most likely due to the crummy weather. We observe a fair array of plains game, lion tracks, elephant tracks, a couple of rhino, baboons and a giraffe. The highlight of the tour is a pair of young male cheetahs with a freshly killed gemsbok calf. We get spectacular photos and video from quite close to the cats.

The ride in the open touring car is chilly, but not bad. We are close to being back at the Welcome Center by 5 pm. We are still in a high-fenced (and double-fenced) area that contains large dangerous animals. Signs clearly state that you may not get out of your vehicle because of said danger. This inspires me to grab Dan's hat, the classic wide-brimmed Akubra, and fling it like a Frisbee into the roadside brush. Now Doc will need to stop the car and see if Dan has big enough balls to get out of the car, brave the lions and retrieve his hat. Doc stops. Dan instead, grabs MY hat and flings it out too. Hunter rises to the occasion and grabs Ross's knit cap and flings IT off into the brush. We all rescue our head gear without being eaten by lions and hop back into the car for the short ride that marks the end of our scenic tour of Kwandwe.

It was good that we were able to use the rainy day to just be tourists. We have not lost any hunting time due to weather, the next day is sunny and glorious. John says that bushbuck rams will be standing out in the sunshine, soaking up a few rays. We head to an old homestead nearby on the conservancy. John has in his pocket what he was told were the keys for the gates. They might be, but they don't work on the current lock on the main gate. John, Nahoy, Spud and I walk in, past the crumbling old farm buildings and up the river valley. It's just as well. It's a narrow little valley and we can see sunny openings that SHOULD have happy bushbuck warming up in them (but don't)! I doubt we would have seen more game had we driven in.

John is a smart rascal and a veritable treasure trove of African outdoor lore. He knows my penchant for soaking up as much local natural history as I am able.

My Name is Tumkulu!

Springbok are the national mammal of South Africa. Fun to watch and hunt. Great table fare.

Thus he is constantly pointing out interesting flora and fauna to me. But he also is not above the odd practical joke in this regard. We are pushing through some thick brush when he hauls up and points at a fuzzy red form. "That is a cow!" he proclaims. Recall that Spud is with us. Ol' Spud is on olfactory overload; both his little nostrils and his feet are moving at many rpm's over redline! He can hardly contain himself. Becoming aware of the cow, he zips toward her, only to find that the old cow has a tiny calf and is rather overprotective of the newborn. The cow turns and comes for Spud. Realizing he is severely out-weighed and outgunned, and not having ANY plan for dealing with something that chases HIM, Spud turns tail and runs right back to John and me. The cow comes boiling out of the brush in an obvious attempt to dissuade the dog from any closer inspection of her progeny. John grabs Spud and finds a convenient tree to duck behind. I don't have a convenient tree, or in fact anything cow-proof. Best I can do is pick up a stout limb lying on the ground. As the cow reaches me, I pirouette as neatly as the most highly trained Spanish matador... and hit the red cow a lick across her slobbery snout! Luckily this works. The cow shakes her head a few times and returns to her calf. I let my intrepid PH know it's OK to come out; the client is safe! Nahoy is nowhere

My Name is Tumkulu!

Wart hogs are the most common of the African wild pigs. Finding one with great ivory can be challenging.

close by; ultimately he was the smart one. He beat feet and found a whole new piece of real estate to exist on!

We head back to the farm for lunch, partaking of a delightful kudu goulash and salad. After lunch we head back out to the old homestead property on the Konapp River. Still no proper key for the gate, so we enter again on foot. At first we see nothing but warthogs and kudu. After the turnaround, we are sneaking back downstream on the side near the old barns when we spot a pretty good common (grey) duiker grazing. John evaluates and proclaims him to be a nice male. I had not planned on taking a duiker this hunt, but it is a trophy of opportunity and has a low trophy fee. I get the .300 WM up on the sticks and make the shot. The duiker drops in place. Nahoy packs the duiker and we trundle back down to the gate where the bakkie is parked. No bushbuck rams are spotted all afternoon, including on the drive back to the lodge.

My Name is Tumkulu!

I enjoy a few more leisurely days of hunting on and around Waterfall Farm. I don't take any more game but I have fun tagging along with Ross and Dan and Frank and Hunter. Now it is time to make an side-expedition down to the coast near Port Alfred. Ross and Dan want to hunt caracal and I will be looking for a nice bushbuck ram. We leave that morning at 5 am, with all five hunters and four bakkies travelling through Grahamstown, Bathurst and into Port Alfred. We meet with "cat" specialist Jeff who will be in charge of the caracal hunt. Jeff explains that his pack of hounds is chasing a caracal as we speak! Suddenly Jeff gets the call that the dogs have treed. It takes us 10 minutes to drive to the jump off. We rush in to the sound of the treed up hounds. Soon Dan has his caracal! Dan has hardly had time to worry about it and he has his trophy cat. We skin it for a full mount.

Jeff takes off with his dog handler. This young man is wearing the South African version of our iconic X-tra Tuf breakup boots. We never learn his name; we just refer to him as Gumboot Boy. Jeff and Gumboot Boy will be looking for a second cat, this time for Ross. Jeff has pointed out a nice little hill on this farm from which John, Nahoy and I can glass for bushbuck. He gives us interesting supplemental instructions though. "This little valley is good. But don't go higher than about where that red cow is. Some Greenpeace people own the property on the top of the hill and they aren't keen on hunting!" Accompanying us is Jeff's son Dwayne, a nice young fellow about 20 years old. Dwayne helps us spot some oribi and several bushbuck. I end up taking a fine bushbuck ram. The most fun is watching John get hung up in a low fence that turns out to feature an electrical wire at just about crotch height! Not particularly funny if you are John, but hysterical to the rest of us.

Over the next couple of days John and I ease back on the throttle. I get to do some cull work, taking a couple of warthogs, a black wildebeest cow and a blesbok ewe to supplement the larder. We all have some fine trophies in the salt. John and Gary haul us back to PE for the flight to Jo-berg and then home, again in First Class, on British Airways.

Chapter 14

Tanana Flats Death March

I did quite a bit of trapping before I came to Alaska. In fact, I had caught what I considered quite an impressive array of furbearers, some in fair numbers. When I started trapping in Alaska, I was well-convinced I was pretty hot stuff in the trapping department. It did not take long for me to conclude I was sorely mistaken in this regard and in fact, I had a lot to learn. When I "came into the country," (which is how we refer generally to immigrants from the Lower 48…) there was no Alaska Trappers Association and no "trapper education" program. There was no "Introduction to Trapping" class at the University of Alaska. All these eventually came along, but at that time they did not exist. So how was a young fellow supposed to learn how to trap new and different species, especially under the extremely adverse weather conditions found in Interior Alaska?

Clearly I needed a mentor or two. But, oddly, not every crusty old trapper (and especially the more successful ones) was willing to jump at the chance of having a rookie tag along to mess things up. This was particularly true if that rookie's stated intention was to "steal his trapping secrets!" I had quite a challenge endearing myself to experienced trappers.

The first such trapper I bamboozled into letting me accompany him on his line, was Ron Long of Fairbanks. Ron had come north in the 1950's from Colorado and married Elaine Evans, a lovely Athabascan lady from Rampart. Ron worked for the Bureau of Land Management, Division of Fire Control (later Alaska Fire Service) in the summer. In the winter he trapped on the Tanana Flats, south of Fairbanks and also bought fur from local trappers to sell at auction. I spent inordinate amounts of time drinking coffee at Ron and Elaine's kitchen table, absorbing trapping lore and tales and salvaging trapping tips of the trade.

Tanana Flats Death March

You bundle up and hope it's enough to do the trick!

Bless his heart, Ron allowed me to tag along on a couple of short trips on his Tanana Flats trapline, primarily to observe him making wolf sets. Ron was picking off a wolf or two each trip from a pack that was close to town and preying on domestic dogs. He called them the "Poodle Pack." I learned a ton from these jaunts and I was grateful. Eventually Ron came to trust me enough that he allowed me to spend more and more time on his line with him. But payback was coming...

Ron had several loops of "line" radiating from his "Birches Cabin" just south of the Clear Creek Butte. He also had a cabin further to the southwest, on the upper Tatlanika River. He reasoned that with another cabin, south of the Birches cabin, he could make it to the Wood River easily enough on the old Bonnifield Trail. Years ago, the Bonnifield Trail had been a major winter road connecting Fairbanks to the mining areas around Wood River, Gold King, and the Tatlanika. On a map in a warm kitchen, it sure looked doable

Tanana Flats Death March

to break out and brush out this line. After all, back in the day, the Bonnifield Trail had been a major travel route. It had been constructed using bull dozers. We should be able to find it and brush it out, at least wide enough for a dog team or a snowmachine. Having a line cabin in the last big spruce stands along the Bonnifield would be the perfect stop over.

Following the Bonnifield Trail south from Clear Creek Buttes was pretty simple for the first six or seven miles or so. Ron was already trapping most of this part of the trail. But at some point, we just flat ran out of marked and brushed "trail." We strapped on our snowshoes and put on some miles. We could not discern where the trail went from there. We spent a LOT of time trying to find it, before Ron gave up and hired a local pilot for a low level flight over the area in a Super Cub. As is often the case, the old trail showed up pretty darned well from the air. Ron chunked out some lathe with survey tape on it to mark the route.

We came back on the ground and managed to find some of the lathe sticking up out of the snow. But there was more good news. On that same Cub flight, Ron noticed that slightly east of the trail there were several open "fields" (in summer they would actually be swamps) through which we could proceed pretty much in a southerly direction without having to cut any brush to establish our new trail.

As luck would have it, the previous summer the Air Force had been doing a lot of work constructing their new Blair Lakes Bombing Range for live fire bombing and strafing practice. Trappers on the flats were generally opposed to the idea, but as is normally the case, the government listened to the concerns expressed by the public and then just did as they pleased. The bombing range was constructed. The Air Force began firing live ammunition on it. And just as trappers had predicted, the whole thing caught on fire. At some point BLM had the Army air lift in some dozers to build a fire suppression line that extended from the impact area, west all the way to the Wood River. Or as we sometimes refer to it...... new trapline!

We reasoned that if we just used our compasses to head due south through the big open fields, we would inevitably intersect the fire line. We also

Tanana Flats Death March

The cabin along the Bonnifield Trail that Ron Long and Howard Luke built.

reasoned that with all the black spruce trees and tundra bermed up along the fireline, there should be some pretty good marten trapping to be had. Marten love piles of jack-strawed pecker-pole timber both for hunting voles and for establishing den sites. Part of our grand theory was right on the money. There were a ton of marten frolicking in those berms. Getting there (and back) proved a tad problematic.

The 1975 Christmas season saw more than the normal compliment of snow to Interior Alaska. Even Ron's regular lines were pretty well covered in deep, fresh snow. Ron wanted to get the existing lines "broken out and packed down." He also wanted to pioneer south to the fireline and "make a few marten sets." He called and asked if I would come along and help break trail. If we could get as far as the fireline, he wanted to make some marten sets to "try out" the area. Does a marten poop in the black spruce forest? Hell yes I would help out! For a few gallons of gas and some of my labor, I would get to watch a master marten trapper at work.

Tanana Flats Death March

We left Fairbanks one crisp January day, crossing the Tanana River at the south end of Peger Road, immediately picking up the northernmost extent of the Bonnifield Trail. Ron was driving six dogs, Elaine had an old Skidoo Olympic and I drove my 1972 single-cylinder Skidoo Elan. We crossed both branches of Salchaket Slough with only minor overflow issues and proceeded about 20 miles to the Birches cabin. Elaine rested up, then headed back north to the river and town on the Olympic. Ron had another machine cached at this cabin for later use.

Ron and I spent the night at the Birches cabin, then took off to the south in the morning. The snow turned out to be a bit much for my little Elan, so I was now breaking trail with Ron's larger machine. Ron followed me on the dogsled, driving the six dogs. On the "weld-a-sled" (Foldasled) behind me, I had camp supplies, including a small wall tent, wood-burning stove, cots, sleeping bags and food for a few days. The plan was to go as far as we could on the previously broken out portion of the Bonnifield, then set up the tent camp. Based from there we would use the snowmachine to break out a trail. Once that new trail was set, we could also use the dogs to haul marten trapping gear.

We set up the tent in a stand of nice white spruce with plenty of dry wood around. I got the stove going. Ron cut and split some spruce firewood and collected spruce boughs for beds for the dogs. The tent was warm and toasty and we were quite comfortable. We spent an enjoyable evening, except for the part where Ron had forgotten to bring any flatware with which to cook and eat! My first task was to whittle a couple of "spoons" out of slices of spruce.

The next morning we made a fateful decision. The dogs were still tired from travelling an unpacked trail the day before. We decided to give them another day of rest. We took off, two of us with just one snowmachine. In retrospect this was not a terrific idea. I have never done it again since! OK, well maybe once.... You can read about it in the chapter about my muskox hunt from Kaktovik!

We easily traversed the open meadows with no significant problems. Ron would leave me with the sled full of marten trapping gear and break trail for a mile or so. Then he would come back, hook up and pull the sled, with me on it, to the next stopping point. Using this method, we made it to the fireline by early afternoon. Instead of doing the smart thing and turning around to go home, we turned EAST (past all the military Keep Out signs and toward the actual impact area!) It wasn't so much the Christopher Columbus gene kicking in; it was more like there was a TON of marten sign. We set a lot of traps. In fact, we set until we were almost out of traps and it was getting dark. We probably would have gone farther, but the Fickle Finger of Fate was pointing directly at us. With an expensive sounding "crunch" the snowmachine lurched to a stop. The engine was running, but there was no power to the track. Kaput!

We flipped the machine up on its side, made a cursory examination, and soon found the culprit. A broken axle at the drive sprocket. We had spark plugs and a few tools, but we did not even have to check our pockets to know that we did not have an axle with us! We were dead in the water. Or, if not in the water, at least in the ass-deep snow.

We first tried to "splint" the axle with a piece of fire-hardened dry spruce wood. It worked..... for about 20 feet. We resigned ourselves to a long walk back to our camp. It was dark now, 3:45 pm. We had been running under mechanical power since first light. We had a long way to go to get to camp.

We could have just made a fire and waited for daylight, but that would not gain us much. Plus, we had left six sled dogs tied up at camp. They would be hungry and, should one get loose, they would probably start eating each other! We had to walk.

We took all our left over snacks (some Pilot Bread, peanut butter, a couple of candy bars and some dried fish). We took what was left of our water bottles. Walking in bunny boots and a parka is a losing proposition to begin with. Even at -30 degrees, you still sweat a lot when walking. Within a mile or two, our parkas were sweat-soaked. We tied them on strings and dragged them behind us. It seemed that time stood still, but the changing locations of stars and the moon in the sky let us know the hours were passing.

Tanana Flats Death March

My cabin on five acres at the Wood River upstream from Wood River Buttes. Built in the mid-80's.

And so we walked on. We got back to the tent around 2:30 am and got a fire going, fed and watered the dogs, had a snack and warm drinks and tumbled into the bed.

We slept until about 9 am when it began getting light out. We left the camp set up and hung the remaining marten gear from stobs on a big spruce tree. Then we loaded our sleeping bags, a bit of food and my fat ass onto the dog sled. Ron jumped on the runners and we mushed back to the Birches cabin where we had the second snowmachine. We pulled into the cabin around dark, stayed overnight, and made it back to Fairbanks the following day.

With a look at the map (again in the warm kitchen...) we plotted our Midnight Death March. We had walked somewhere on the order of 26 miles at 25 to 30 degrees below zero. Neither of us wanted to do it again!

Tanana Flats Death March

This is a Tanana Flats wolf that I trapped from the Poodle Pack.

A couple weeks later, Ron and I went back to the tent camp in the big white spruce. This time we also had Athabaskan elder Howard Luke with us. We stayed in the tent again, picked up the marten traps we had set prior to the Death March, and cut logs for a line cabin at the site where the tent was set up. Ron and I came back to town. Howard stayed and worked on building the line cabin.

Last time I looked (from the air) the cabin was still there. Apparently the military has not found it yet. The Army and the Bureau of Land Management (BLM) have a policy of "deconstructing" cabins that they find on lands included in the Public Land Order (PLO) on the Tanana Flats. Ron was murdered in 2003 by a demented trapping partner. Elaine passed away in 2014. Howard died in the fall of 2019 at the age of 96. It could very well be that I am the only living soul (not incarcerated...!) who knows the location of that cabin.

Chapter 15

Hunting In the Land Of The Hobbits

New Zealand has become a classic international hunting destination. There are no large game that are indigenous. All the interesting creatures that are classified as game in New Zealand have been introduced over the years. There are no large predators, so human hunting is the only control on populations. At times man has been inefficient in his control efforts and some species have become pests. But there is a plethora of great game animals there and hunting opportunities abound.

I have hunted New Zealand twice. Once on my own, as a side-trip on my way to Australia. And the second time, this narrative, with my friend the late Bill Miller Sr. of Llano, TX. As best I recall, Bill had never before hunted anywhere outside the U.S. It was a great trip; his enthusiasm was infectious.

Once again I am flying First Class on Alaska Airlines "miles." This time on Qantas. And once again, flying in decadent luxury with the extravagant elite folks who actually can afford it makes the flight an integral part of the adventure instead of an ordeal that one tolerates to reach one's destination.

After crossing the Pacific in pampered luxury, I deplane in Auckland. It's early morning, pre-dawn. Ordinarily I would have no problems in Customs, but my friend Lynn has given me a wooden toy to bring along to give to Ross Ferguson when I get to Australia. Any "wood items" must be declared, and thus I am in the "problem child" line. This line does not move as rapidly as the "Nothing to Declare" line.

The small world thing being what it is, the Agricultural Agent on duty is a Maori fellow who has visited Alaska and had a terrific time touring our state. As it turns out, among other places, he visited the Simon Paneak Memorial

Hunting In The Land Of Hobbits

Rangetiki River from the ridge out from the lodge.

Museum at Anaktuvuk Pass; not many tourists have done this. I am his first customer this morning and we have a nice chat. He inspects my hunting boots for errant soil, finding none, because I washed the Vibram soles before I left. He also enjoys playing with the carved wooden moose that Lynn has sent to Ross. The little toy moose actually poops black jelly beans when the neck is levered.

I pass through and ease my way out into the terminal, finding a currency exchange and turning US $100 into NZ $158. I also find a "free phone" and call my motel. The shuttle arrives promptly, driven by a nice guy from Fiji who has a brother working on a floating fish processor in Alaskan waters! It's the small world thing again...

At daybreak, I go outside and walk the neighborhood. It is in an industrial area, sort of an air park. But I find some shops offering food within a mile. I have 24 hours to kill while I wait for Bill Miller to show up. I check my

Hunting In The Land Of Hobbits

email, visit the exercise room and at sundown, go back over and procure some "pub fare" in the form of fish and chips.

The next morning I pop out of bed, have breakfast and check out at 6 am. There is no sign of Bill anywhere in the domestic terminal, but I check my bag with Air New Zealand and walk back over to the international terminal. No sign of him here either. Since his flight has been on the ground for two hours, I assume I just missed him at the other terminal and walk back. This time I find him. We have a grand reunion, then BS for a couple hours waiting for our flight to Palmerston North. The flight goes smoothly and Colin is waiting for us just inside the terminal. Poor Colin, he is tasked with guiding both of us! Colin was my guide the last time I visited this hunting establishment. We got along well and thus I requested his services again.

It is a really neat two-and-a-half hour drive from Palmerston North up to the lodge. Once there, I reunite with a lot of friends among the staff. I first hunted here a couple of years ago. It is a "full house" in terms of hunters. Bill and I have some lunch and then head down to the range to sight in. Bless his heart, the outfitter at my request, has procured from somewhere a .50 caliber in-line muzzleloader. I can't wait to try it out. But hang on... there is in fact a teensy problem. The rifle is so filthy I can barely force a projectile down the dirty bore. It appears to have NEVER been cleaned! Nor do they have a cleaning jag. As one might imagine, the "sighting-in" does not go very well. The best I can do with the rifle in its current condition is a 20 inch group at 100 yards. I may be reduced to using the .308 that Bill has borrowed if I cannot figure out a way to clean the incredibly filthy muzzleloader.

After our range session Colin takes us on a tour of some of the property in a side-by-side ATV. Bill is fast catching on as to why I told him I thought three days was plenty of time for him to take a good red stag. We see a lot of deer on our tour, including tons of reds, elk, sika, fallow and even a rusa. Bill spots several reds that he proclaims are "plenty big enough" for him! One in particular is heavy and wide. A really nice rack with no stickers or other "trash" at all. Bill wants to stay out all night and try and keep an eye on this stag. Colin assures him we will have a really good chance of finding him close to this same spot tomorrow morning. The sun goes down, the air turns chilly, and we head back to the lodge.

Hunting In The Land Of Hobbits

Alpine Hunting Lodge viewed from on the mountain.

We meet the large crew of other hunters as they come in from their jaunts of the day. They are from Louisiana, North Carolina and South Carolina. Being from Texas, Bill can understand their Southern drawls. Turns out that the Louisiana contingent is a family group who are alligator farmers! Interesting, gregarious bunch to be sure.

After supper I retire to the shop in the garage to tackle the problem of the dirty muzzleloader. Colin is interested and willing to help, but it is not an easy task. We are lacking the normal compliment of cleaning and take-down tools. I am handicapped by not being able to take the rifle apart and extract the breech plug, much less completely submerge the barrel. As we wildland fire fighters are wont to say, I will "do the best I can with the limited resources available to me!" A good dose of field expediency is required. With the barrel still completely attached to the stock, I use a 20 gauge copper bore brush and some homemade patches to swab the bore with good old fashioned dish soap and very hot water. Finally I give it a light coating of gun oil and hope

Hunting In The Land Of Hobbits

Bill Miller with his beautiful big red stag.

for the best. We made amazing progress, but it took over 2 hours. I remain nervous about not being able to extract and clean the breech plug, but at least we have given the barrel a decent scrub.

The next morning dawns cloudy and 43 degrees. Bill and I are up at 5:45 am. It does not take much to get Bill out of bed and launched toward the kitchen and dining room. I think he stayed dressed and awake all night thinking about the red stag that Colin showed him last night. We file into the dining area for a self-serve breakfast. I will head out for the day stoked up on English muffins smothered with local honey, some sliced fruit and half an avocado. Colin, Bill and I depart about 7 am, over to Slaughter Valley. On the way we see red deer, fallow deer, sika deer, feral goats and Arapawa sheep. We glass from several vantage points, then head back to "First Lookout" to try and relocate Bill's chosen stag that we saw last night while scouting from the "Honeycomb." Bingo! There he is...

Hunting In The Land Of Hobbits

We dash down off the lookout and across the valley toward where we spotted the stag. After a short climb through the manuka scrub, we top out on a little ridge within 130 yards of the last place we saw the stag. The big deer is still there. To me, he looks bigger than the "Silver" medal size that Bill is contracted for, but I stay uncharacteristically silent. What do I know about judging red stags? I've taken exactly one! Colin glasses the stag thoroughly and gives Bill the OK to shoot. Bill has settled in prone. He is solid and apparently had all the slack out of the trigger! Colin's verbal instructions are still floating on the morning air when the .308 goes off, hitting the deer high in the lungs and wasting no meat. The stag is dead but doesn't know it, so Bill shoots again. At the second report, the big stag flips over into a brush patch and lies completely still. Graveyard dead. After some back-slapping and hand-shaking, Colin heads out for the ATV. Bill and I walk down to the stag.

The smile on Bill's face is proof he now actually believes me that three days of hunting on this property would give him "plenty of time" to take the stag of his dreams. There had been some minor disbelief on his part, but it is all gone now! It is the morning of Day One and Bill has his trophy! Back at the lodge for coffee and a snack by 10:30 am. After that brief interlude, I go out to the skinning shed and give Colin a hand with caping and butchering. Then the meat is placed in the walk-in cooler. On a normal hunt, we would be racing right back out to hunt again, but there is no shortage of game here and Colin deserves a nice lengthy break.

After lunch Colin and I head back to the range with the now much cleaner inline muzzle-loading rifle. In its improved condition, the rifle now shoots a 2 ½" group at 100 yards off the bench. My efforts at cleaning the bore have obviously paid off. I was afraid I might have to resort to using a modern center-fire rifle if I could not get the muzzleloader to group well enough to be used responsibly as a hunting rifle.

The project for the remainder of the day is to see if we can scare up a Silver Medal fallow stag for me. We head to some of the highest elevations on the property, a good 1,500 feet higher than the lodge. It's a fun expedition; I get

Hunting In The Land Of Hobbits

some great photos of red and sika deer. We have a look at the new "outpost" camp well above timberline. It has a million dollar view for sure. It's cold and windy up here this afternoon. We see only a few sika stags on the way up and virtually nothing in the higher country. From another lookout on the way down, we watch one of the other hunters stalk and kill a magnificent stag. Even back at the lodge I never hear what the score turned out to be, but I did catch the price tag. I can assure anyone who will listen it was above the budget of either Bill or myself! WAY over! I'm glad the hunter is happy.

At supper, it becomes clear we are the only blue-collar hunters in this crowd. One of the husband-wife pairs have been gone from the property today; they are back for supper, whereupon we learn their mission today was "looking for a winery" for James to buy for Sheila for her birthday!

After supper Colin puts the tape on Bill's antlers. The SCI green score is 332, which is actually several inches above the top limit of a Silver Medal stag. Technically this is a Gold Medal animal. It was the guide's call, so there will be no surcharge. Bill has cinched himself a good deal indeed.

It's cloudy and overcast the following day. We pop out of bed at 5:30 am and are out the door by 7:30 am. There really is no pressure on either of us any longer. I know I won't have any trouble finding a good fallow today or tomorrow. I am extremely happy Bill has taken a really good red stag. The rest is just a lark of an adventure in really cool country! The three of us pile into the ATV and head down the entrance road, through a cattle guard and up the far side of "First Lookout." We see several fallow stags; there is no shortage. None are more appealing to me than a light colored stag that we saw earlier this morning. We head back down to try and relocate him.

Colin and I spot the stag almost simultaneously. It is time to begin the stalk. For some reason, I invite Bill to come along. I should have left him in the ATV! We sneak to within 149 yards. If I had my own muzzle-loading rifle, I would try this shot. But I am nervous about shooting at this range with a dirty, borrowed muzzleloader. I let Colin know I'd prefer to get much closer. This decision has consequences, since there are two more fallow and a

Hunting In The Land Of Hobbits

sika stag between us and our quarry. These other deer will likely function as "lookouts." The fallow stag I want to try for is grazing contentedly. Thus we patiently await further developments.

After 15 uneventful minutes, one of the other two fallow stags begins walking our way. This stag is quite a bit larger than the stag we set out to stalk. I'm sure Colin won't let me shoot him, since my contract is only for a Silver Medal fallow. So Colin and I haul up and remain motionless. Bill is several yards off to the side. He cannot see the close stag, only the one grazing unaware of our presence some 150 yards away.

To the extent I can do so without alarming the closer stag, I am trying desperately to get Bill's attention to shut him down. But Bill is preoccupied. He takes off his cap and examines it. He scratches his bald head and adjusts his sleeves and vest. I am trying valiantly to get his attention and stop him from fidgeting around. Finally Bill spots the closer fallow, but by now, the deer has him firmly locked in his sight. The stag is paying rapt attention to the man-creature from Texas that fidgets about!

I am still standing up. The stag has us pegged and I have no chance to take a kneeling or sitting position or even to deploy the shooting sticks. It will be an off-hand shot or nothing. Colin has his binoculars up and is studying the stag. This deer is a borderline case. Colin of course wants me to get a deer at the top of the Silver category. It is his duty to not allow me to shoot a deer some high roller would pay big money to kill. Colin's reputation is on the line; there is more pressure on him than there is on me! He needs the deer to turn so he can properly assess the palmation, but the stag's attention is riveted on us. We have only a head-on view. I think he is still trying to figure out what kind of weird Smurf Dance Bill Miller was doing.

The stag takes a couple of steps in our direction and turns to his left, at what later turns out to be 48 yards. He's sure something is wrong, but perhaps he is not convinced we pose any deadly threat. He is apparently more worried about the Smurf Dance guy than he is about me with the rifle. The stag hesitates once more and Colin hisses "Crack him...!" I am ready and before

Hunting In The Land Of Hobbits

Arapawa sheep were introduced to New Zealand many years ago by explorers who wanted to have a meat source if they returned.

the words have left his mouth, the rifle goes off. The stag wobbles 20 yards and goes down. I was sure enough he was hit solidly that I violate my own First Rule of Hunting with a Muzzleloader: "Always reload immediately...!" We all walk over. The stag has been killed cleanly.

The good news (or bad, depending on your perspective) is that Colin has again misjudged the trophy. The deer's antlers are quite a bit bigger than he judged them to be. This means the stag is bigger than the Silver Medal class animal that I have contracted for. Just like Bill, I now have a Gold Medal stag for a Silver Medal price tag. Stuff happens! Colin isn't upset, so I guess I'm not upset either. We make our way back to the barn and skinning shed. We cape the deer, hang the meat in the cooler and munch on sausage and cheese for lunch. After lunch I clean the muzzleloader while Bill takes a nap.

Soon we are back out on the mountain for an afternoon photo safari. I have talked myself into shooting an Arapawa ram if we see a good one. We aren't

Hunting In The Land Of Hobbits

30 minutes from the house when Colin spots a band of sheep. One is a real corker, but Colin talks me down! "Let's go look at some others." We proceed all the way down to the Rangitiki River. Ever the natural history buff, Colin points out some of the fun, light-colored igneous rocks at the river that actually float. And we see some nice rainbow trout in the one to two pound class. After a while we leave the river and start back in the general direction of the lodge. Cresting the pass we almost immediately spot the band of sheep with the big, black ram. We have not seen a better ram; Colin gives me the go-ahead to try for him. I slip off to the side and work on getting close enough for a good shot with the muzzleloader. The terrain is hilly and broken; there is little skill involved in getting in pretty close. At around 50 yards I haul up, set up my shooting sticks and touch off a shot. The flock of sheep take off for parts unknown. The big ram lags behind and tips over. I have my "big, fuzzy sheep!" Back to the skinning shed we go.

While Colin is caping the ram, Bill and I spot a young hedgehog in the garden. A new fun pet to play with! We race over and take some photos. I finally figure a way to pick the thing up. It's sort of like a porcupine, but the quills are not nearly as sharp. Nor is he nearly as aggressive as a wild porcupine would be. When I pick him up, he curls into a tight little ball, quills out. Turns out this is no ordinary hedgehog. He has been painted with gold spray paint and stands out amongst other hedgehogs! "Why?" you may ask. The staff here have a running hedgehog wager going on. They take the "gold" hedgehog out a mile, or two, or five and dump him out. He knows when he gets back to the house, he will get cat food as a treat. The guides, cook and others place bets on how long it will take him to come back! He is back now and someone will have won the money in the pool. Cheap thrills in hunting camp.

Today, April 25, is Anzac Day. It is the New Zealand and Australian version of our Memorial Day and pays tribute to veterans. It is an important holiday in both countries. We arise at our regular time, but have a relaxed breakfast. The high rollers are going to be hitting it hard today. They have several trophies left on their "want" lists. We let the rest of them head out to comb the hills. On the other hand, Bill and I are all done killing things. Our day

Hunting In The Land Of Hobbits

This is the guilded hedge hog that kept the lodge staff entertained.

will feature photos, fun and food! We head west to a box blind. We sling out some maize (shelled whole field corn) and have a seat in the musty old blind. The structure has not seen any use in "a while." The seat is busted and from the looks of the interior, birds have been living (and crapping....) in it. The interior is pretty well coated with bird dookie. We sit a while, but no game shows. We choose to head up toward Windy Pass. On the way up we stop and peer into the deep, heavily vegetated canyons and draws. We see a variety of beautiful reds, sika, fallow and even a couple of rusa. After getting some great photos, we head back around 10 am.

At the skinning shed, I tally the trophies from just this week's hunt. The high rollers have been busy! There are six red stags, two sika stags, four fallow, four Arapawa rams, one stinking feral goat, two tahr, one chamois and a single NZ possum! Quite the haul. The high rollers have tallied up a substantial bill for trophy fees. They have all been shooting "big stuff," not the mid-level "Silver Medal" trophies that Bill and I pursued. After lunch,

Hunting In The Land Of Hobbits

My trophy fallow stag, taken with .50 caliber muzzleloader.

we head out on another photo tour. I get great pictures of several reds, rams and goats. Supper in the lodge is complete with a special blessing and a toast hoisted by guide David, himself a veteran. Davy lives on a neighboring sheep station and guides here at times.

The next morning Bill and I hop aboard the van and are hauled back to Palmerston North to catch our in-country flight to Auckland. There Bill and I part company. Bill will return to the States and I proceed on to Australia for another adventure with Dan Field at Narooma, New South Wales.

<u>Note:</u> I have very special fond memories of this expedition with my Texas friend. Sadly this was Bill Miller's last hunting adventure. Within weeks he passed on. When his red deer antlers finally arrived in Texas, they were picked up by his son, my friend Bill Miller Jr. They now adorn his wall. A fitting tribute to what I suspect was the culmination of Bill Senior's long hunting career.

Chapter 16

Hunting Dinosaurs

For years outdoorsmen have listened to stories of the fabulous annual outdoor show held each February at Harrisburg, PA. Never before had I had a chance to actually attend. But in early 2009 a friend from New York talked me into going to Harrisburg with him for the show. It was an interesting experience. Huge pavilions, thousands of people, hundreds of outfitters and purveyors of outdoor gear. There were a number of Alaska guides exhibiting who were friends of mine. I enjoyed walking around and seeing the exhibits. I loved talking to the outfitters. I had no intention of actually booking a hunt, but at some point, I came to the conclusion there really were some good deals. Deals so good, it seemed, it would have been a crying shame to pass them up. My resolve weakened...!

There are few guided hunts even in the Lower 48 states, where you can hunt for a few days for less than two or three thousand dollars. At this show, there were a number of outfitters peddling alligator hunts in the swamps of Louisiana and Florida. Most seemed to be priced at between $2,500 and $3,000 for a decent sized 'gator. I noted however, that nearly all of the 'gator outfitters had a LOT of openings. They were not selling many 'gator hunts. To me, this seemed like an opportunity to wheel and deal. I ended up getting a pretty good deal on a $2,500 gator hunt. Our final handshake was on $1,300 for a good chance to take an 8 foot alligator. AND the outfitter would throw in a couple feral pigs! I was hooked. I wrote the deposit check and continued wandering around the show!

It is in September that I actually head to south Florida to chase the wily alligator. Upon landing in Orlando I collect my duffle bag and pick up a rental car. Thanks to Walt Disney, traffic is terrible, but I finally make my way out of Orlando proper. I proceed south with a stop at McDonalds and finally find a $38 motel room. Between my crappy map and GPS I plot a route to Lorida, FL for the next day, and hit the hay.

Hunting Dinosaurs

Around lunchtime I stop and call the camp. The outfitter's wife asks me not to show up quite yet; there are hunters from the last bunch still in camp. To kill time, I stop and have a leisurely lunch at Sebring. Sebring is not a sleepy town today. There are between five and ten thousand motorcycles in town for the Southern equivalent to the Sturgis, South Dakota (SD), Motorcycle Rally, event. After a bit I slip out of town. I finally find Lorida, the ranch, then the camp. All by 2 pm.

My outfitter, Danny, drives up. I introduce myself and he shows me to my very comfortable accommodations for the next few days. I relax sitting in a rocking chair on the front porch while Danny goes back out to make arrangements for the rest of his incoming hunters. In the south, porch sitting is a time-honored tradition that I could get used to.

From my porch perch, I suddenly hear a tremendous ruckus. Along come the yard dogs. They look to be a cross between Jack Russell terriers and beagles. They are headed toward the river, hot on the heels of a 150 pound black feral hog! The whole yammering crew heads off through the palmettos and other underbrush. Never a dull moment.

After supper Danny explains that we will be hunting for my alligator at night, along the Kissimmee River, on public land. His favorite spot is a series of canals that are currently under the control of the US Army Corps of Engineers (COE) and a contractor. The COE has decided in their infinite bureaucratic wisdom that the dredging they encouraged years ago to drain swamps was actually a bad idea. So now they are using more taxpayer dollars to put the dredged material BACK into the canals! The canals are now classified as a construction zone. The public is not allowed to boat in the area, even at night when no one is working. This all apparently makes perfect sense to the COE bureaucrats, just not to normal people. Danny knows they are serious about the closure though. A friend of his was cited for going through the area in an airboat a few nights ago. We will not be hunting there.

In preparation for my hunt, Danny went up to the Avon Park Bombing Range office yesterday and bought a $57 permit to gain recreational access to the Range. The permit includes use of their boat launch. The permit allows

Hunting Dinosaurs

Danny and his family access for one year. He can add "guests" to the permit for $7 each. By launching from that particular site, he can save a 30 minute airboat run compared to launching from his camp and running up the river to the same hunting spot.

We leave camp with the airboat trailered, at 6:30 pm. We arrive at the Avon Park gate at 7:15 pm. Here a cranky, overly-officious female security guard announces huffily that "no airboats are allowed!" One would think this critical tidbit of information would have been useful to pass along yesterday when Danny bought the permit. Since we are on Sergeant Tightass's shit list already and she obviously does not approve of airboats or the societal misfits who own and use them, not surprisingly he gets taken to task for yet another tiny infraction. Since Danny had planned to use multiple vehicles for launching over the course of the season, he has affixed the $57 decal to a piece of glass that can be moved from vehicle to vehicle. She announces that Danny will be getting an "official letter" admonishing him for having his decal on the piece of glass instead of "permanently affixed" to the windshield.

Danny prudently doesn't argue with the cranky, officious woman. Young Mikey, the driver is not quite as smart. He smart mouths the guard on a couple of occasions, but we are able to turn around and get back out to the road unscathed. We head back to camp, having wasted an hour and a half. At about 9 pm we are launching into the river within a few yards of where we had supper.

Once we get out onto the marsh, Danny stops the boat and explains to me how this adventure is supposed to work. First, his headlamp, mounted on his cap, is the only illumination we will have. Whatever direction Danny looks, is where the light will shine. Next, I'm handed a cross-bow and shown how to operate the safety and the red dot scope. The cross-bow is loaded with a stout bolt tipped with a heavy-duty harpoon type fish arrow. The head is trailed by heavy braided line, almost like halibut ganion line. Danny explains that he will spot the gators and try and judge their size. This professional judgment is based on the distance between the red, glowing eyes highlighted by the headlamp.

Hunting Dinosaurs

The porch with rocking chair was my favorite spot (aside from the dining area!)

The airboat is noisy; normal conversation will be futile. This is made more evident by the fact we are all wearing hearing protection in the form of heavy duty ear muffs! So, if the 'gator is a big one and fair game, Danny will tap me once on the shoulder. The second time he taps me, I am supposed to shoot. Further, I am to aim at the 'gator's neck. Although a hit almost anywhere will work, a solid hookup in the neck is best in terms of "reeling in" the 'gator so it can be killed with a "bang stick." The "bang stick" is a PVC pole with a .44 Mag cartridge in a small metal cylinder at the end. When "loaded," one taps the gator on the head, releasing the firing pin and discharging the cartridge downward. I have requested that I be able to salvage the gator's skull, so care will be taken to "tap" the gator just behind the skull, severing the spine but not destroying the skull. After my five minutes of intensive instruction, we are off.

Hunting Dinosaurs

This thermometer was a constant reminder that I was NOT home in Fairbanks.

I immediately realize what piece of gear I have forgotten. My glasses. There are tons of insects flying around the Florida swamps this warm fall evening. If I do not keep my eyes nearly closed, many of the little rascals are ending up in my eyes. Not that glasses would have helped a ton. Other insects are getting into my mouth and nose. I can shut my mouth and avoid entrance through that orifice, but breathing seems somewhat critical. The basic sensation is not pleasant. On the bright side however, we are seeing a TON of alligators! I don't think we go more than 30 to 45 seconds at a time without spotting a pair of gator eyes reflected by the light. Most of these giant reptiles are in the three to five foot range, but a few are larger.

Finally a big one, in a good location, is spotted. Danny taps my shoulder, I am ready. I aim the red dot scope on the crossbow for the neck and rise to the occasion, despite my limited training and experience and lack of spectacles. There is a huge splash and then nothing. Mikey shuts off the boat. Danny

Hunting Dinosaurs

proclaims I missed and suggests that I reel in the line and have a look at the arrow and head. I first grab the plastic buoy, then the ganion line and begin reeling the contraption in and coiling it in the bucket again. Danny explains that the arrow will no doubt be stuck in the mud. I will need to tug fairly hard to get it free. When I've reeled in all the slack and come up against the stuck arrow, I give it a healthy tug. As it turns out, I have not missed after all. I am now "playing" a thoroughly pissed off alligator on 30 feet of sash cord. With no gloves or other hand protection I might add. It is an interesting task to say the least. Rough on my dainty paws!

The 'gator has thrashed out a bit further from the boat. I am pulling the line in, hand over hand, but the gator is not in any sense heading our way. In fact, he is motoring strongly toward the mass of woody vegetation along the shore on the far side of the slough. "Don't let him get in there" yells Danny. The harder I pull, the stronger the gator swims for shore. Since I'm not giving up any line, he is basically pulling the whole boat with him.

Considering how many 'gators I have seen, this bugger looks huge to me! He has made it to a mass of vegetation and tangled the line. He is hung up. We close in and I get a 10 second crash course in how to safely and efficiently use a bang stick. I arm the cartridge and three different times I smack the gator just behind the skull as directed. And three times the shell fails to fire. Each time I crack him with the stick, the gator gets a little more agitated. But he doesn't get dead!

Danny grabs the bang stick and extracts the cartridge. The primer is dented, but apparently the pin is not hitting hard enough to ignite said primer. Thus the continual misfires. Danny puts in another cartridge. Apparently I have failed this portion of my test. Danny may be concerned I will figure out how to get the stick to fire while it is pointed at my foot or at the floor of the boat! Danny hands the bang stick to Mikey. Mikey cracks the gator a good lick one time and promptly blows a large bloody hole in the top of the gator's skull. So much for "please save me the skull!" But then again, I must admit he killed the gator graveyard dead and things on board this airboat are a lot calmer than they have been for the last five minutes.

Hunting Dinosaurs

My alligator trophy well on his way to becoming a pair of boots and some great meals.

Mikey hauls the gator in over the gunwale of the boat. This thing is a very impressive beast to be sure. The locking plastic tag is affixed. We are now all legal. We motor back to the camp landing, offload the 'gator and then move gear and 'gator into the truck. Mikey and Danny will store my gator in the walk-in cooler. My gator hunt is over, nearly as quickly as it started. I'm back in my cabin before 10:30pm. Danny stops by to ask if I want to go pig hunting in the morning. Do gators poop in the swamp? Hell yes, I want to go pig hunting. Danny says "Be on the porch, ready to go, at 5:45 am."

It's 61 degrees, calm and overcast the next morning. I leap out of bed about 10 minutes before the alarm goes off at 5 am. I walk over to the kitchen, have something to eat and head back to my cabin. We head out for a few hours, but find no piggies. Still, I'm happy my gator hunt has gone smoothly, albeit quickly. At this stage in my career, I'm a sucker for just "being in camp." Thus I happily hang out, in between unsuccessful forays for pigs. Part of my

Hunting Dinosaurs

"lack of success" is due to the fact that any pig I shoot is not going to tip the scales at any more than about 35-50 pounds. I am in this pig procurement project for the meat. I especially do not want some stinking big boar with large tusks, like all the other hunters crave!

Sure, I am the proud owner of a dead alligator. But now I have a whole new problem. Like a farm dog who has been chasing trucks for a few years and finally caught one..... What do I do with it?

We ponder this dilemma while we take some "hero shots" of my prize. Danny gets the gator out of the 'fridge and hangs it from a rafter in the barn. We take a few photos. Then we discuss how to handle the trophy. Danny will skin the beast and butcher up the tail meat. He will flash freeze the meat so I can fetch it back to Alaska with me. The salted hide will go to a tannery in Georgia that specializes in tanning alligators. While the hide is being processed, I will have plenty of time to decide what I will have made from it.

At some point I decide on having a pair of custom made western boots created from my Florida 'gator. The job goes to Loveless Custom Boot Company in Oklahoma City, OK. On one of my frequent deer hunting trips to Oklahoma, I stop in at the Loveless store and get measured for boots. I'm lucky enough to bump into the company's owner, Bob Loveless, who happens to be in the store that day. We have a grand conversation and I am delighted to learn that Bob is a hunting buddy of my Congressional representative, Don Young of Ft. Yukon, AK, "Congressman for All Alaska!" It is still a small world.

Since I had plenty of alligator skin for a pair of boots, I also have Bob craft a nice western belt that matches my fancy new boots. With the leftovers from all this activity, I also prevail on my friend Mountain Man Mark to make me a matching holster for my Leatherman tool.

These days I wear my custom made boots and matching belt and holster when I go to a dinner auction or fund-raiser. I always stick a snapshot of me and my gator in my pocket. In that way, I have proof of my claim. I am normally the only guy in the room who "shot my own boots...!"

Chapter 17

Mekoryuk Muskox

It took me several years to "get over" my first muskox hunt. It was not so much the near death experience on the ice of the Arctic Ocean that kept me out of the muskox hunting business. It was more the embarrassment of having proved that I, an experienced Alaska hunter and, perish the thought, a Master Guide… had not been able to distinguish from a distance between the girl muskox and the boy muskox! I was, euphemistically speaking, sort of "snake-bit". I was not anxious to dive back into the challenge of hunting for a muskox.

Being a big believer in the old adage that "when you get bucked off, you must get back on the horse that bucked you off," I spent some time speaking about my indiscretions (and subsequent embarrassment) with the single most knowledgeable muskox hunter in Alaska. The late Edward Shavings Sr. of Mekoryuk, on Nunivak Island, was undisputedly that authority. Ed, a Cup'ig Native elder, was born and raised on Nunivak. As a boy, Ed watched the original introduction of muskox to the island circa 1935 when the animals were first brought to Alaska from Greenland. He watched the population grow and flourish and he participated in the first hunting season on the island in 1975. Ed was one of only a handful of Native licensed guides. He ultimately became a Master Guide. He guided on Nunivak even well into his 80s. He was one of the smartest and hardest-working individuals I have ever met. Ed and his wife Esther were dear, dear friends. Not only did Ed sweep my fears and hesitation aside, he offered to help me do a "re-take" of my muskox quest. I decided to take him up on it.

It's now early March 1993. I had flown from Fairbanks to Anchorage to attend the spring Guide Board meeting. The meeting is over, but the Iditarod is due to leave town today. I get up early and drive downtown with the intention of talking with some of my musher friends as they get ready to leave for Nome in the race. I meant only to be a spectator, but I end up working as a handler for my friend Sonny Lindner of Johnson River.

Mekoryuk Muskox

Ed Shavings Sr holding court in a shelter cabin on the south side of Nunivak.

The following day I pull myself together, turn in the rental car and catch a flight to Bethel. ADFG technician Brad meets me and another incoming muskox hunter. Brad ferries us to his office for our orientation class, a condition of our hunt permit. I pay particularly close attention to the part about how to tell the bulls from the cows! As with my GMU 26 registration permit a few years ago, this permit is for a "Bull Only." Once we are through with our class, Brad drops us at our respective B&B's. Only in rural Alaska would a breezy 30 year old ATCO unit be considered suitable quarters for tourism!

Weather the next morning looks quite flyable. I catch a ride to the airport and check in at Era Aviation where everything goes smoothly except the baggage allowance is 40 pounds and I have 116 pounds. This will be a little expensive, but not nearly as bad as what is going to happen when I need

Mekoryuk Muskox

to fly off the island with a couple coolers of frozen muskox meat and a cape and skull! These items obviously, are the objective of this adventure. By the time we are ready to board, the weather has deteriorated to some degree. The pilot explains that instead of going directly to Mekoryuk, we might first go to Tooksook Bay and wait for the clouds and fog on Nunivak to dissipate. As we get airborne the clouds lift and we have smooth sailing, landing at Mekoryuk right on schedule. Ed meets me with a snowmachine and sled. Soon we are sailing over the drifts to his warm, comfortable home in the village.

Upon reaching Ed and Esther's house, I find two old friends there. Brian, a Girdwood guide and Rick, a Fairbanks bowhunter, are both there, hunting with Ed. They had traveled to the southeast corner of the island yesterday. Both have taken fine musk ox bulls. The guys are not scheduled to go out yet. Their meat and skins are being worked by Ed's crew over at his son J.R.'s house. We head over there and work at packing all the meat and trophies into wet-lock boxes for air freight shipment back to Anchorage.

With this chore completed, Rick, Brian and I head over to the area where the reindeer herding and butchering took place in the fall. Brian shot a couple of white (Arctic) fox there a few days ago. The foxes are feeding on the residuals from the butchering. Brian says there are a ton of foxes there. We spend a couple hours and get a few shots, but it turns out there are a lot of red fox and very few white fox. After three hours of walking around the drifts, we head back to Ed's house. Foxless, but it was better than sitting around the house watching RATNet television!

Light wind, overcast, snowing and 15 degrees the next morning. Rick and Brian are scheduled to fly back to the mainland. Ed's final client for the season, an Anchorage surgeon, is due in on today's flight. The Era Aviation agent calls to say the passengers will go, but the freight cannot fly today. We unload the hides, capes, horns and meat boxes. Just the personal duffle and weapons will accompany the passengers.

Assistant Guides JR and Sam begin preparations for hunting later today. I'm staying out of the way; I had no idea we would take off so late in the day. But I'm flexible and willing to do whatever Ed decides.

Mekoryuk Muskox

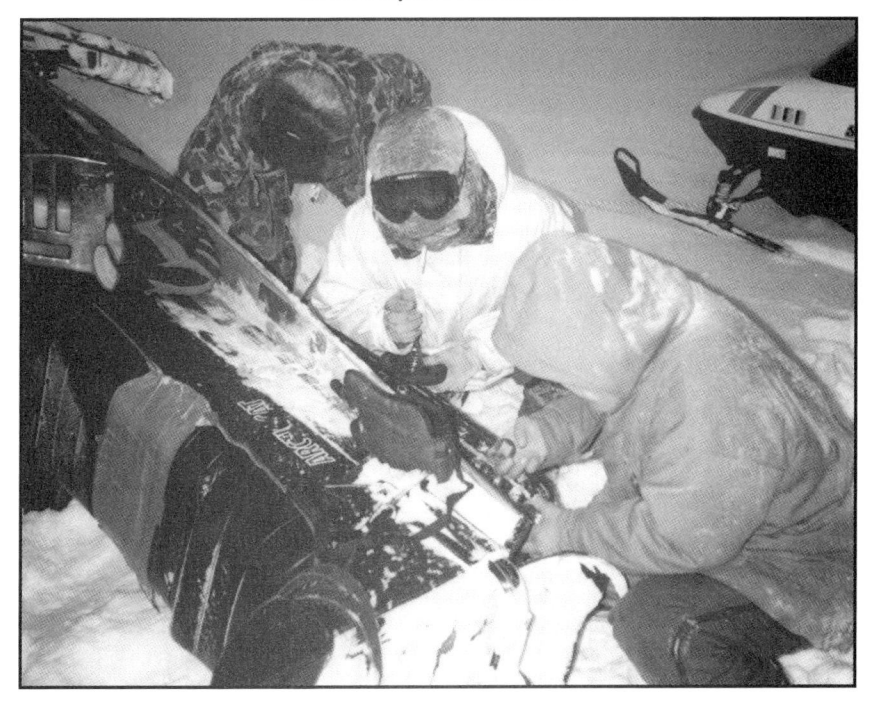

Ed and the boys replacing a bearing on the drive axle. Apex resourcefulness!

Just as Brian and Rick are getting on the sled to head to the airstrip, the Era agent calls back: Era Aviation will indeed accept freight today! We go through another mad scramble to reload the boxes and get them back over to the airport. I stay in my safe zone: in the kitchen talking with Esther.

Ed hauls the guys and their mountain of gear to the airstrip. JR gases up the snowmachines. Ed is soon back with the new client, Randy. Turns out Randy is a maxillofacial surgeon. There's a relief. If we go over a cliff in a snowstorm and I land on my face, good medical care will be close at hand and I will be well taken care of.

Our little procession heads out by 11 am. Off we go, to the southeast. It seems like a nasty blowing snowstorm to me, but to the denizens of this barren island in the Bering Sea, apparently this is good enough weather to

Mekoryuk Muskox

hunt! I'm riding a komatuk sled being pulled behind JR's Arctic Cat. In addition to the falling and blowing snow, bits of compacted snow and ice are picked up by the machine's tracks and slung up into my face. As much fun as it would be to watch the countryside, I finally opt to pull a piece of an old sleeping bag up over my goggles and trapper hat to protect my pretty face. We run southeast for about two hours.

In the marginal visibility, Ed has led us a bit farther to the East than he originally planned. During breaks in the wind and ground blizzard, we find ourselves between Nanwaksjiak Crater (a long extinct volcano) and Twin Mountain, near the old Ingrimiut fish camp. Armed with glimpses of familiar terrain features, Ed reorients himself and we turn slightly southwest toward Cape Mendenhall. The cape was our original destination. We can see Ingriruk Hill near the Nanathloogagamiutbingoi Dunes. We can all see the dunes, but only the Yup'ik speaking members of the party can pronounce them! Caucasian lips simply cannot form the sounds needed to pull that off! From Ingriruk Hill we observe the Fish and Wildlife Service cabin on Duchikthluk Bay. We swing over to the cabin to get out of the wind, warm up a bit and have a mug-up. Wildlife is becoming more plentiful as we near the south coast of Nunivak. We have seen a lot of reindeer. Some alive and some winter-killed. The carcasses are well-attended by both white and red foxes. In the "fox world," finding a reindeer carcass must be like winning the lottery. "Whoopee! We gonna make it 'til spring!"

After our brief warm-up session in the shelter cabin, we circle Ingrijoak Hill and travel the outer beach of Cape Mendenhall. For the first hour, we see no muskox. We keep heading northwest along the upland side of the Bangookthleet Dunes. As soon as we get into the dunes, we begin seeing muskox. Buckets of them! In no time at all, we have looked at about 30 nice bulls. I'm getting antsy to kill something. Ed is of course, far more stoic. He is "Yup'ik-silent" as per usual, but I can tell that the wheels are turning. I have all the faith in my Eskimo friend, but I still want to shoot a bull muskox!

Finally we locate a lone bull in some small dunes. We leave the machines well back and circle in. I stick close to Ed as he sneaks in closer for a better look. We have previously discussed that I would prefer an older bull with wide, heavy bosses. If a bull fits that description but does not feature the widely acclaimed "black tips" that show on perfect, un-broomed horns, so be it. I would rather have the "old man of the dunes" when it comes to my *ommingmak*.

We ease in to within 100 yards of the bull. One moment Ed is studying the bull intently through his binoculars. The next moment he is telling me to shoot. I carefully chamber a 175 grain Nosler round in my trusty H&R Ultra Rifle 7 mm Rem Mag that has been with me for more than 20 hunting seasons. It came into my hands from my dear old dad, as a graduation present and for finally making the Dean's List (instead of "academic probation") at forestry college.

Before heading out on this hunt, I had again visited the Large Animal Research Station at the University of Alaska's Institute of Arctic Biology. I got to run my hands over some real live muskox to determine how their physiology worked. Some muskox guard hair can be over 20 inches long and the location of the precise point of aim can be problematic. Remembering this information, I hold behind the line of the front leg, and as high on the shoulder as the animal's eye. I squeeze off the round and feel good about the hit, but the bull does not go down. I have purposefully avoided hitting bone, in an attempt to minimize meat damage. Ed urges me to shoot again. I hold for the same spot and lower the boom again. The bull disappears down the far side of the dune.

JR, Sam and the other hunter catch up to us and we walk together over the dune. As we top out, we can see the bull, all his life memberships cancelled, just over the crest. He is a magnificent old creature. He has only a hint of black on one horn tip, with all the black broomed off on the other. This configuration gives the impression the horns are shorter than the un-broomed bulls that Jeff and Rick had taken. But when Ed slaps the tape measure on my bull, we find that he is almost 1 ¾ inches longer than the others. This bull's horns are just massive, with huge bosses. I am plenty happy, but watching Ed and the boys discuss the situation in Yup'ik, it becomes clear I have not been fully appreciative of what I have here.

Mekoryuk Muskox

We proceed with the obligatory photo taking, then I drop back out of the way to allow Sam and JR to dive in with blades a-flashing. The speed they demonstrate at reducing a dead muskox to a sled-load of meat, hide and horns is wondrous. I'm in awe. This is obviously not the first muskox they have sub-divided. In less than an hour, we are moving again, looking for a good bull for Randy.

The sun actually pokes out for a bit as we follow a line of small, rocky hills to the northwest. As we enter the Bangookbit Dunes, we again begin seeing little bands of muskox. We look over another 20 or 30 bulls before finding a group of three big bulls that Ed decides are worthy of a closer look. There are two older animals with broomed tips and one younger bull with perfect black tips. JR and Sam circle around. Ed, Randy and I, also on foot, wallow directly up the seaward side of the hill.

Ed proves once again that he is indeed the "Muskox Whisperer!" We come out just 10 yards from the three bulls. We are seeing them from the rear and the side. Ed wants a frontal view, so he grunts and gets their attention. They immediately back up to a large drift and give us the wild muskox traditional "you are now too close" look. Their defiant snorts have got me feeling a bit mortal at just 30 feet away, but Ed examines them patiently. No need for binoculars either...

Ed finally decides which bull is the best and tells Randy to shoot. Randy is ready and lowers the boom with his .300 Win Mag. He admits that when he relates the story, it will be "a running shot at 300 yards" instead of standing still, broadside at 30 yards. Only those of us hunkered down in that particular snow drift will know the truth: at 10 yards, he could have about killed it with a sling shot or a dull Buck knife!

With the large bull down, the other two hang around only momentarily, then move off along the top of the dunes. We assess the situation. The consensus is that this not the best place on the island to skin and butcher the animal. First, it's on the tippy top of a hill where we cannot easily get the snowmachines and sleds. Second, there is a lot of sand blown into the snow and no clean place to lay out the quarters to cool and set up. We huff and we

Mekoryuk Muskox

My muskox. Down, out and ready to be converted into quarters of meat and a fine hide.

puff and pull the bull to the side of the hill and roll him over the edge. From the base of the dune, it is a short way to drag him onto a clean patch of snow. I take some photos of Randy and his quarry. Our intrepid guides walk back a half mile to bring up the snowmachines and komatuks.

The bull is caped, skinned, butchered and boned by 6:30 pm. After a 15 minute break for rest, snacks and drinks, we are saddled up and ready to head back into Mekoryuk. Ed and the boys study the mountain tops and volcano craters that we can see to the north in the distance. When they are sure where the village lies, they take a compass heading. Each machine has a compass mounted on it for travel after dark or in white-outs. That will be the only way to navigate tonight. Note: GPS's were not in common use at this time. Sleds fully loaded and then some, we take off at a brisk 20-30 mph pace across a big lagoon and north into the gathering darkness.

Mekoryuk Muskox

Loading the ERA Twin Otter at the Mekoryuk airstrip, getting ready to fly back to Bethel on the mainland.

After two hours of fairly uneventful (just cold as hell...) running, Ed's machine blows out a bearing and a rear bogey wheel flies off of his Arctic Cat. The part is critical. It is one of the pair that keep the track aligned and under tension. Regardless, Ed doesn't seem to notice or even slow down!

JR scoops the bogey wheel up, one handed, as we pass by Ed and his sled at 25 mph. A short time later, Ed seems to be sensing the difference in how the machine is performing. He stops to examine the suspension as JR drives up with what is left of the bogey wheel. The bearing is completely shot. Sam and Randy soon catch up to our mechanically induced cluster. After a short conference in Yup'ik, a mechanical repair plan is formulated. We two English-speaking Caucasians don't actually get to participate (or even eavesdrop) on the discussion. Not that there is much a forester or a surgeon could do to help!

Mekoryuk Muskox

I am about to be reminded of just how damned resourceful our Native brethren in rural Alaska are. Remember we are not working in a warm shop somewhere. We are outdoors, in the dark, at 0 degrees F, in a 15 mph wind. With the machine propped up on its side, Sam begins the process by pouring hot coffee on the nut that secures the bearing and wheel onto the axle. This thaws the nut enough that he is able to get it turned and backed off.

He then attempts to remove what remains of the old bearing from the axle. The immediate problem turns out to be how to press the old bearing out of its casing. The guys glance around in the dark. No one can see a bearing press laying on the tundra! I do my best to function as a "clamp" to hold the wheel onto the ski spring mount, while JR attacks it with a screw driver and an axe. After 10 minutes of intense but rather fruitless effort, we pour more hot coffee on the unit on the chance that it is both frozen and tight. Even that does not do the trick. We move the entire operation over onto the ski suspension system of JR's machine. Finally the faulty bearing is pulled off. JR and Ed quickly press on a new bearing that mysteriously appears from Sam's repair kit. Then we bolt the wheel back on and tighten the track tension. We are back in business and ready to take off. Ed calls Esther on the CB and tells her we are about an hour out. Off we go again, in pitch black of night, with only the snowmachine headlights for guidance.

At Ikathiwik Crater, Ed calls again and estimates 45 minutes out from the village. From the crater, we can see the strobe on the Mekoryuk airstrip, as well as a few faint lights in house windows. We wind our way down off the last hill and onto the ice of the Mekoryuk River. In just about exactly the estimated 45 minutes, we pull up at the house where JR lives. Here we off-load the capes, hides and heads. Then on to Ed's house where Esther has an incredible muskox pot roast waiting. It's after 10 pm and we are some hungry hunters.

In bed by 11 pm; what a great day and a grand adventure. I am reminded today of the great respect I hold for my Alaska Native friends, their way of life and their resourcefulness on this rugged remote landscape that they proudly call home.

Chapter 18

Rams On Ice

My first mountain hunt in Alaska was memorable. It was 1974 and I was to pursue Dall sheep, much like the intrepid nimrods I had dreamed of and read about in Jack O'Conner's stories in Outdoor Life back in the 1950s. I was still in the U.S. Army at the time. My boss, Captain Bob Bean, had been the intelligence officer for the 4th Battalion 9th Infantry, 172nd Arctic Light Infantry Brigade when I was first assigned to Fort Wainwright. Bob had taken a couple of Dall rams during his time stationed in Alaska. I was a Sergeant; mere "common enlisted swine," but Bob either took a shine to me, or just felt sorry for me. He did me the honor of inviting me along on a sheep hunt in the fall of 1974. In fact, over the 4th of July holiday weekend, we hiked more than 20 miles up the Johnson River from the Alaska Highway and cached some food and fuel in the last spruce trees below the moraine of the Johnson Glacier. The plan was that we would not have to pack in as much weight when we actually embarked on our hunt in late August.

Ordinarily, the location of one's Alaska sheep hunting spot is kept a secret as closely guarded as news of a case of an STD at a convent. But since this hunt took place nearly 50 years ago and the area now is open only to holders of lottery drawing permits, I have no compunction about revealing it in some detail. It remains to this day, a good place to hunt sheep. Because of terrain and ice, it is a fairly difficult area to hunt.

By the time our hunt was to begin, Bob had been reassigned from Ft. Wainwright to Fort Greely, Alaska, 110 miles down the road from Fairbanks. Among other duties, Bob was assigned to a high elevation rescue unit. As luck would have it, when it came time for us to start our hunt, he was obligated to "standing by" as the only qualified commander for that unit. Thus our hunt was put off a couple of days. We decided to make up the lost time by getting flown in, rather than hiking all the way up to the Johnson Glacier from the Alaska Highway.

Rams On Ice

These Dall ewes passed close to us in the limited visibility of the snowstorm. When they caught our scent they quickly left.

I drive to Delta Junction, meet up with Bob and get ready to fly into the Alaska Range. Our pilot, Roy, is out flying a search-and-rescue mission. We finally make contact with Roy around 6 pm. I stuff my lanky frame (I've always been tall; in my 20s I was tall, but not fat....) into his Super Cub and take off. We fly up the Little Gerstle drainage, behind Independent Ridge and then up to Johnson Glacier. On our short jaunt, I spot over 150 sheep, including some decent looking rams. This is gonna be great! I am all in a dither over being on my first mountain hunt.

Roy lands me at Porter Strip, on the east side of the Johnson River. We can only hunt on the west side. The Alaska Board of Game recently designated the Tok Trophy Management Area for the area to the east. We applied, but did not draw permits. There is an airstrip at the moraine of Johnson Glacier

Rams On Ice

on the west side, but there is a guide camp there. Big Delta guide Charley Boyd and Roy have an agreement. Roy won't fly anyone else into the strip until Charley is finished hunting for the year. Fair enough. (Note: Years later, after Charley's death, I was awarded exclusive use of this area by the Guide Licensing & Control Board. I established my main camp at this same air strip and spent many a fine day hunting and guiding from here for 30 years. My late wife Jan's ashes are spread there at the old camp on the moraine.)

Roy drops me off and takes off to go get Bob at Allen Army Airfield on Ft. Greely. I don't wait for Bob. Per our plan, I hike on upstream about a mile and choose a fine flat, dry spot for our first night's campsite. I set my heavy pack down and cut some willow sticks for use as tent stakes. Five minutes later I return to my pack. Bad news! Clever fellow that I am, I set my pack down right smack dab on top of a whole bunch of highly teed off ground-nesting yellow jackets. There are hundreds of yellow jackets buzzing about and they are as friendly as a dad who found out his daughter is going to the prom with a registered sex offender.

I cannot even get close enough to my pack to grab it and race away. I'm nearly out of options, but I finally settle on starting a little smudge fire a few feet upwind of the cranky insects. I feed it with green willow brush and alder leaves and after five or 10 minutes the yellow-jackets are groggy enough I can safely retrieve my gear without getting stung. I select yet another flat, dry spot that is not having a special on stinging insects. Here I am able to set up the tent, get a decent fire going, coffee made AND get back down to Porter Strip in time to meet Bob as Roy lands just at dark. I help Bob pack his gear up to our new and improved (yellow jacket-free) campsite. We eat supper and crawl into our sleeping bags, happy to finally be in the mountains, albeit still on the "wrong" side of the Johnson River for our hunt.

Up and at 'em at 5 am. We are following the faint track of John Porter's old horse trail heading upriver by 6 am. We cross Prospect Creek and later, Elting Creek before 8 am. Both side-canyons are excellent sheep country, but they are on the "permit" side of the river. We need to reach the "beginning" of the Johnson River, at the terminus of the glacier's moraine. Our chances of getting across the Johnson River down here without drowning are roughly

Rams On Ice

the same as me being named Playmate of the Year. Climbing onto the moraine is no picnic either. After our creek crossings and frontal assault on the moraine, we are both wet from our boots to our waist. I'm in worse shape than Bob, having fallen in during the last crossing. My 50 pound pack made swimming somewhat problematic.

We take plenty of time reconnoitering a safe route across the moraine. Once we are onto solid ice, the going gets easier, but we attach crampons to our boots to hopefully preclude sliding into the odd glacial crevasse. We hike all day in light rain. By early evening, we reach our predetermined base camp location, at about 5,500 feet elevation, just off and above the marginal moraine on the northwest side. We are far above the vegetation line now, so all cooking must be done on a small, gasoline mountain stove. The wind is picking up and changing direction. Crawling into a sleeping bag seldom felt this good.

The following morning we sleep until almost 8 am. I don't think either of us realized how worn out we were from fighting the high-water creek crossings and getting onto and across the moraine, not to mention meandering up the glacier itself. We packed over some rough ground yesterday. The secret to good hunting in Alaska is always being willing to do more work than the next hunter. This morning the sun is shining brightly. We take this opportunity to spread out our soggy gear on the rocks in the sun to dry. The brisk down-glacier breeze, coupled with low relative humidity means drying our clothes and gear does not take long at all.

While the gear is drying, we wander about clad only in our long johns and do the requisite caching of extra food and supplies. We will take food and fuel for three days and press on higher. The remainder of our gear and food supplies go into a rock cairn that hopefully will prove relatively creature-proof. We harbor no illusions that if discovered, a determined grizzly or even a wolverine could put us into a bit of a pickle.

We also put some time into cleaning the rifles; they have had a rough, wet trip so far. I borrowed a .270 Winchester from a neighbor. I'm of course still enthralled with Jack O'Conner and Jack has long preached that the

Rams On Ice

.270 Win is the "best sheep rifle." It's in pretty good shape for the trip it has had. Bob's .300 Win Mag, on the other hand, obviously has issues. Water and glacial silt are crammed into it. The muzzle is blocked; it is not safe to fire in this condition. We do not have a cleaning rod. Bob's brilliant "idea du jour" is to take a long wire with a cleaning patch affixed to the end and pull it through. From the muttering Bob is uttering, apparently this is not going smoothly. Although he cannot cuss loudly (for fear of spooking any nearby sheep, we communicate only in whispers now...) he is making some seriously upset noises. I walk over to observe and kibitz. Bob has jammed the slightly over-sized patch about half-way up the barrel. In the course of trying to pull it through, the wire has broken. Without a rigid cleaning rod (something we clearly should have brought and didn't...) the rifle is useless. Deader than the concept of free lunch. We are down to a single rifle for the two of us. There is nothing we can do to resurrect his .300 Win Mag. We cache the inoperative firearm with the spare groceries. Bob wants to shoot the .270 for "familiarization." We chance spooking sheep, but it probably would be best to confirm for sure that it is still sighted in. A couple of shots is all it takes. Bob is now introduced to and "familiar" with the .270. We are both convinced it is still zeroed. Hopefully we will find our rams in different bands and at different times!

Our plan today is to climb to between 7,000 and 8,000 feet, then to slowly move, parallel to the fall of the glacier, while glassing down for sheep between our route and the glacier. We leave camp at 10:30 am. By 1 pm we are at approximately 7,400 feet. From our perch we observe over 50 sheep! There are some legal rams (legal ram in this case is a ¾ curl ram) but no really large sheep. Since we can always "settle" for a young ram, we keep moving. Hopefully toward where the big ones are!

Around 2 pm, the wind shifts abruptly 180 degrees. It begins to snow. Flurries at first, then some fairly serious snow. We push on, but the snow comes down harder. We assume it will stop soon. After all, it is only mid-August! We find a little rock ledge and tuck in out of the wind and blizzard conditions to wait for the weather to clear. Visibility is now down to well under a quarter of a mile. It really is not wise to continue moving along the mountain under these conditions.

Rams On Ice

I wanted to peak around UNDER the Johnson Glacier. Beautiful, but a little creepy!

We doze for a bit. At one point a Dall ewe passes within 50 feet of our spot. The poor dear almost has a heart attack when she catches our wind! At 4:30 pm, it is still snowing heavily and the white stuff is staying where it has fallen on the ground. Clearly we will be spending the night here on the mountain. Descending is simply not a safe option.

The only place even remotely "flat" upon which to pitch a tent is a little pass on a hogback between two side drainages that has two sheep beds on it, side by side. So this is where we put up our four foot by eight foot mountain tent. It's no easy feat to get the tent up in the wind coming off the mountain. We finally get it erected in spite of the wind, which we estimate is now gusting 30 or 35 mph. We rig an extra three lines of cord on the windward side to "storm lash" it. Once the tent is up we crawl inside and polish off our first day's allotted rations, of the three day's worth we are packing with us. The sleeping bags are

Rams On Ice

well-wrapped in their waterproof covers. They are completely dry and thus nice and warm. It is hard to sleep due to the noise of the storm raging around us and battering the tent. Neither of us sleep for more than an hour or so at a time. Toward dawn, the wind drops off. But this presents a new problem. Without the wind, the snow sticks to the top of the fly on the tent. When it accumulates enough snow, the soggy tent starts sagging down on top of us.

Now it is Day Two of our three days of rations. We get up at 8 am and crawl out of the tent. It is still snowing; there is a foot or so on the ground. This is no weather to travel in. We have no water left in our canteens and must melt snow to provide water. Hopefully we are packing enough gas to run the stove to the extent we will need to get the water we need. After melting snow for water to drink and draining ourselves off a bit, we get back in the sleeping bags. We nap on and off throughout the day, still tent-bound.

The next day we poke out of the tent about 8 am. It is still snowing lightly when we crawl outside, but the storm seems to be slacking off. The wind has swung back around 180 degrees and high above us there are patches of blue sky! We hold hope it is clearing. We eat breakfast, such as it is, and begin breaking camp. After all the time we have spent in the tent, a lot of condensation has formed and frozen on the tent and the fly. The result is that it takes a major operation to fold the tent and fly small enough to tie onto our packs. Each stiff fold risks a tear in the nylon fabric. We finish packing and start off the mountain very carefully in the fresh snow. Visibility is better, but not good. We must be very careful not to inadvertently march over a cliff or drop-off concealed by drifting snow and marginal visibility.

In less than an hour we have descended to the marginal moraine and a nice flat area. By 1 pm the storm has dissipated. It was moved out by a southerly down-glacier wind. Not only has it stopped snowing, but there is actually some sunshine poking through the storm clouds. We have a spectacular view of Mt. Gakona, snow encrusted and shining magnificently, across the glacier from us. We get our new camp all set up and celebrate with a nap. While we really don't expect to see sheep this soon after the storm, Bob goes out with the spotting scope about 5 pm. I stay in the sack. Bob is back within just a few minutes. He is shaking me and whispering loudly the news that there are

Rams On Ice

two rams in the bowl above our new camp. One ram is a legal ¾ curl. Bob has killed other sheep in the past and still has higher standards. I, on the other hand, am a complete neophyte and have much lower standards. Considering the weather and our diminishing food supply, I would like to try for the legal ram. Low standards are better than NO standards!

We watch the rams for 30 minutes and figure we have two possible routes for a stalk. One is very high and entails about a half mile hike in full view of the sheep. We end up discarding this option, mostly because of the obvious lack of opportunity for concealment. Instead, we go down over the marginal moraine for about a mile, then up the back of the spine the sheep are on. There will be some wind issues to chance, but we have a prevailing down glacier breeze. We should be OK as long as we do not climb to an elevation as high as the sheep are. We head down to the glacier and then begin our climb. By 7:45 pm I've reached where I approximate my best shooting position should be. I crawl to the crest and peek over. I cannot see the smaller ram. The larger one is at about 250 yards, quartering away downhill.

I am fervently hoping my scope has not suffered any alignment issues in our adventure to date. I hold high on the last rib on the near side and squeeze a round off. Down he goes; all his life memberships cancelled. The ram appears to die before hitting the ground. Bob goes for our packs; I head over to the sheep. Halfway to the kill site, I stumble into the small ram. He trots by me at less than 10 yards. Further up the mountain he passes Bob at about the same distance.

We reach my ram and find it is seven years old, between ¾ and 7/8 curl, measuring somewhere in the 34 inch range. We get him skinned and quartered by 8:30 pm and head for camp. We are back at our flat spot by 9:15 pm. I carried the ram off the mountain myself and will sleep better for the effort. It is a very satisfying expenditure of calories! We place the meat, horns and cape in a hastily constructed rock cache to hopefully protect it from grizzlies and from a group of wolverines we spotted earlier. We "wash up" with snow... and pop into our sleeping bags.

The next morning we awake to severely overcast skies. Bob had cautioned me to expect about one decent hunting day out of three in the sheep mountains. That seems to be the norm. With an eye toward staying fairly close to camp in case another storm hits, we decide to just hunt the mountain above camp.

Rams On Ice

My ram, taken immediately post-blizzard and a day or two pre-march-to-the-road. A ¾ curl ram was legal in those days.

Soon we spot a band of six rams and come up with a plan for a stalk. After two hours of hide-and-seek in light snow and fog, we reach our selected shooting position and peek over. The rams are gone. We track them in the 12 inches of snow on the ground, but after three hours, it is clear they are headed for the other side of the mountain at a speed much faster than we humans can muster. We turn around and head back to camp. Our ascent to 9,000 feet today is the highest elevation we will reach during this hunt. We partake of a tasty but skimpy meal and turn in.

Up the next morning to another beautiful sunny day. Great weather. However, we have a looming problem. We are now on Day Five of the three days of food we brought along. Obviously we have sheep meat to supplement these rations, but cooking it over a small gasoline camp stove would be difficult.

Rams On Ice

Also, our fuel is in fairly short supply. The Sheep Hunter's Diet for yesterday and today is as follows: Breakfast consists of three dried dates and a ½ cup of hot chocolate. Lunch is six dried dates and a small piece of caribou sausage. Supper is half a Mountain House freeze-dried meal and a cup of tea. To beat the dehydration issues, we are each trying to drink two to three quarts of water laced with Kool-Aid, each, per day.

Our plan for today is to move up-glacier a bit to the last mountain that has any bare rock (and presumably some grass...) exposed. It lies below a snow/ice field that backs up to the very crest of the Alaska Range and falls away down the other side. On the map it appears to drop off onto the Canwell or Castner Glaciers on the Richardson Highway side. We are now closer to the "Rich" than to the Alaska Highway! We figure it's the last place where a sheep could wrest a living from this icy winter wonderland. This must be where "the Big One" lives...!

This morning we move up glacier by walking on the glacier surface itself again. The hard surface in the center is much easier to move along than trying to negotiate the marginal moraine. We head down the mountain, across the moraine and out onto hard ice. With crampons and ice axes, we can move safely on the ice. We avoid the huge crevasses on the outer edges. About two miles out of camp we spot sheep; four rams. After a thorough examination with the spotting scope, we determine while all are legal, none of them are up to Bob's standards. They are however, smack dab in the middle of the only feasible route up our target mountain. Getting around them unseen is out of the question without more technical gear than we have with us. We decide to just keep moving slowly, making no effort to hide or sneak. Perhaps they will decide we are granola-crunching mountain climbers instead of sheep hunters? It works! They remain uninterested in us, moving slowly up and over the ridge they are grazing on without spooking. We make it to the top by 11 am and set up the scope. We can now see another band of rams. There are seven rams feeding along the edge of the moraine some 2,500 feet below us, about a mile away. There are some interesting horns in this bunch. They bear closer examination; so we move along.

When we are within 700-800 yards, we take another look-see. Bob decides even though none of these rams are spectacular, we are rapidly getting to the point where he should shoot something and then both of us should high-

Rams On Ice

tail it out of this snow-beaten glacier country before someone gets hurt. He decides to make a stalk and try for the largest ram. This will be a complicated stalk, with a lot of the mountainside to be covered, almost completely in the open. Being "Army guys" we have cleverly brought along our government issue "over-white" parkas. Hopefully this camo ruse will work. Only time will tell.

As we palaver and plan the stalk, the rams all bed down like the spokes of a wheel. There are eyes looking in all 360 degrees! I elect to stay at our current perch and watch the show through binoculars and the spotting scope. Bob will do better advancing on his own. He takes the .270 and stealthily heads off toward the band of bedded rams. Within 45 minutes I am pretty well frozen to my rock perch. The plot thickens as the rams arise from their beds and move over the spine of the ridge. Now they can no longer see Bob. But then again, Bob cannot see them either. He can now move much faster, rather than low-crawling through the snow, trusting his over-white parka to conceal him. In 30 minutes he has reached the point from which he had planned to shoot, but there is nothing to shoot AT! Being an experienced sheep hunter, Bob moves forward again very cautiously. He is closer to where they went over the spine when suddenly they begin feeding back this way. Bob studies the rams carefully, then picks up the rifle again. From what I can observe from afar, the sixth ram in line is the largest one. Evidently, Bob thinks so too...

As ram Number Six crests the saddle, Bob appears to fire. Five of the rams all jerk their heads up, then go back to digging for grass in the snow. Then I see ram Number Six fall. Momentarily I hear the report of the rifle. Bob has shot the ram cleanly through the neck; I can see some blood on the cape. The problem of course is that he was aiming for the shoulder! The scope has had a hard trip up the glacier, but all's well that ends well. I grab both packs and make my way over to where Bob and his ram are. It takes me 20 minutes to get across. In sneaky mode, Bob used twice that much time to travel the same route.

The ram is very nice. Its light brown horns measure over 38 inches along the widely flared curve. A well-executed stalk and a lucky shot with a dinged up

Rams On Ice

rifle have resulted in a fine trophy. Bob capes the ram. I bone the meat and get it stashed in our packs. We pack up and are headed back to our "high camp," reaching it by 4 pm. Over what passes for supper, we decide to try and reach Boyd Strip at the terminal moraine of the glacier in two days of hard packing. Our packs will be quite heavy. We will be carrying a whole sheep each. Thus our packs each will weigh more than 100 pounds. We tramp back down the glacier, reaching our objective around 10 pm. This is NOT the recommended way to safely travel in glacier country.

The next morning we sleep in until 8 am. Our goal was to leave by noon and indeed we do. This time we have even heavier packs. We have our supplies and camp gear in addition to a sheep apiece! We move down-glacier in light wind and sunshine and make really good time. We should be able to get to Charley Boyd's camp and strip by nightfall. If we do not, we will have a devil of a time sleeping out here on solid ice. Luck is with us and we pull into Charley's camp just after 9 pm. The camp is right at the tree line and we are delighted to have a crackling, cheery campfire to celebrate and dry out. A nearby creek provides all the drinking water we require. No more melting snow to get water.

We left a fair bit of food at the high base camp this morning. This is a gamble since we are now depending on utilizing the cache we made here near Charley's camp over the 4th of July weekend. We accomplished a 26 mile walk in from the Alaska Highway, stored the stuff in a tree, and walked the 26 miles back out to the highway! For fun! We can already taste that big ol' can of "Big John's Beans and Fixin's!"

As luck would have it, a grizzly has found our cache and pretty much "fixed" it beyond recognition! Most of the cans are scattered on the tundra and gravel. Virtually all of them have at least one grizzly tooth mark in them and thus are spoiled. The bear did not actually eat a lot of groceries, but he sure rendered them inedible for us. We seem to have possibly gotten even to some degree. Mr. Bear has eaten an entire tub of boot grease and some chemical fire starter tablets! And he bit into a can of Blazo fuel. That combination quite possibly caused some gastro-intestinal distress. True confession? I hope it did! Retribution for rendering my "Big John's Beans and Fixin's" inedible!

Rams On Ice

This was what was left of our food/gear cache from the original walk in. Some grizzly was not impressed on how we "secured" it!

When the bear smacked the waterproof bag tied into the spruce tree, stuff flew around enough that several cans caught in the branches of nearby trees. We are able to salvage six cans of assorted food stuffs. We open and consume four of them for dinner. A "grab-bag" supper as it were; all the labels are gone! Peaches and creamed corn never tasted this good; who knew? Plus, now with plenty of wood, we can feast on sheep meat. Fat and happy, we await Roy and his Super Cub, due in tomorrow.

We sit on the strip all day in blue-bird weather. Our stomachs may have shrunk a bit over the past week, but we do our level best to stretch them back out to their former capacity. By 9 pm, darkness is falling and there is no sign of Roy. We hit the hay.

The following day dawns featuring weather that is not conducive to flying. Thus we have a decision to make. We can either wait here and hope the

Rams On Ice

weather clears and the wind drops off. Or we can hike out to the Alaska Highway. With winds of 25 mph, gusting to 40 mph, there is a ton of blowing loess silt in the air from the glacier and river bars. It's not that pleasant just sitting here, so we decide to walk out. This will be 26 miles and it will entail several crossings of side creeks off the main Johnson River. We know the route from having hiked in to cache the food and gear in July.

At Charley's camp, we cache all our sheep meat, horns, capes and some gear. With light packs (a bit of food for snacks, water and rain gear) we head for the road. Mid-morning we take a break of five minutes. Lunch becomes another break. Around 4 pm we stop briefly again. The rest of the time we are moving steadily toward the pavement. I was 23 years old at the time. Today I would not try that hard of a push! We make the Lisa Lake trailhead on the Alaska Highway at 6:15 pm. A long, hard day for sure, but we are out to the road.

At the pavement in a pullout, we meet up with a sheep hunter from the Anchorage area. He has a permit for the newly established Tok Trophy Management Area on the east side of the Johnson River. He knows little about the area, so we trade some intelligence about the terrain, conditions and hunting on that side in exchange for a 50 mile ride back up the Alaska Highway to Delta Junction and Ft. Greely. Bob calls and checks in with our pilot. Then we treat ourselves to dinner at the Evergreen Inn followed by showers and some quality sleep in real beds at Bob's apartment in the Ft. Greely BOQ. We are now new men. Unrecognizable from the dirty zombies who just walked out from the glaciers at the crest of the Alaska Range!

The next day Roy flies back to Charley Boyd's air strip to pick up our gear and our sheep. What a terrific hunt. Successful in terms of adventure and experience. Sheep meat in the freezer and horns for the wall. A few "survival" situations, but no one was injured; we both survived in fine form. As I dreamed and suspected when I read about this in Outdoor Life long ago, I think I'm going to like the challenge of being a sheep hunter in Alaska. In fact, I'm sure of it!

Chapter 19

Kodiak: One Shot Per Day

Number one son, Jason, had graduated from the University of Alaska at Fairbanks in the Fire Science program. He was almost immediately offered a firefighter job by the City of Kodiak Fire Department. Of course he hauled up and moved to Kodiak. Who wouldn't? After a year working for the city, he moved over to the Department of Defense Fire Department at the Coast Guard station in Kodiak. Fantastic! Now I had a convenient "base camp" on Kodiak! No more hefty hotel bills while waiting for the weather to clear so I could fly out! In the summer of 2003 Susan and Jason had presented me with my first granddaughter. Jaden was born there in Kodiak, but I had not seen the tyke. There was no venison in my freezer. Jason had some leave time. The stars were obviously aligned. What better time to fly to the "Big Island" and do a bit of deer hunting?

October 24th and it is 15 degrees and overcast in Fairbanks. Ironically it is a good way to acclimatize to Kodiak weather! I hop on an Alaska Airline flight to Anchorage where I switch planes and fly on to Kodiak. At Kodiak it's cloudy, foggy and raining hard; typical fall weather on Kodiak. Jason and three month old Jaden are at the airport to meet me. We run by the Coast Guard station to see where Jason works and meet the firefighters of B-Shift. A week before, I had sent Jason a grocery list for our trip. I sort of optimistically envisioned him procuring those groceries, but that has not come to pass. On the bright side, at least he still has the list; he has it with him, so we do the grocery shopping. Then we pick up Jason's wife Susan and head out to Henry's Great Alaskan Restaurant for a great supper. At the restaurant, we bump into a number of other firefighters, both City and Coasties. I also see my Alaska State Trooper friend Mo. Mo and I reminisce about the Donnelly Flats Fire in Delta. As the Agency Rep for Alaska Division of Forestry, I was riding in Mo's cruiser with him during the evacuation of Ft. Greely and portions of the outskirts of Delta Junction. At one point I assured Mo we could probably safely proceed past some flames close to the road. When part of his light bar melted, I revised my opinion....!

Kodiak: One Shot Per Day

We even have dinner entertainment, put on by a drunk backing his car into an SUV in the parking lot. By golly, this is the Kodiak I remember and love from when I started guiding on the island back in the 1970s!

Before we head to bed, we make a quick call to Seahawk Air to confirm our charter for tomorrow. Our flight was originally scheduled for 9 am. They are covered up from weather delays and ask us not to show up until noon. I hit the hay in my sleeping bag on an air mattress on Jason's floor. It has been a long travel day.

The next morning the weather is slightly better. It is clearing to the southwest when I get up at 6 am. We hit King's Diner for breakfast and confirm with Seahawk Air that they want us on the dock at 12:30 pm. But once back at the house we get yet another call from Seahawk. "Can you come a little early?" We dash over to take advantage of a break in the weather. We weigh in at the float dock and load the Beaver. Between the two of us, we are allowed 370 pounds; we have just 210. There are two other hunters on the flight, so the fact that we are way under the limit probably helps them. It helps my checkbook to be able to split the charter bill for the flight in with the other two guys.

An hour later we are in the air over the south end of Amook Island. We have flown over the northeast end of Kodiak, past Uganik Lake and through Zachar Pass, across the east end of Spiridon Bay, the east extent of Zachar Bay and then down the west side of Amook in Uyak Bay. The ceiling is low and we never get more than 2,000 feet Above Sea Level. I get to see a lot of country that I slogged around in years ago while guiding bear hunters for Leon Francisco at Zachar Bay and Old Uyak. Despite the scud running, the flight is quite smooth. We land at our transporter, Steele's Amook cabin at 1 pm. It is still cloudy and rainy.

The two other hunters are from Anchorage. They will be staying at this cabin on Amook Island. Steele picks us up with his mini-landing craft and moves Jason and I across Uyak Bay to the old Parks Cannery where we will be staying. We settle into quite comfortable quarters. Steele has partitioned off a portion of an old bunkhouse. It is heated with an ancient Monitor oil

Kodiak: One Shot Per Day

stove and has a large open living/dining area featuring a propane oven and range. All very civilized to be sure. As much time as I have spent on Kodiak in soggy tent camps, this is going to be very enjoyable. We have a wonderful place to get warm and completely dry each night! The lap of luxury.

We toss our gear into our cozy bunk house quarters and opt to take a hike up the mountain behind the cannery buildings. Deer are not covered by Alaska's strict "same day airborne" law, so we can hunt today, the same day we flew in. We start hiking around 4 pm, not really expecting to accomplish much other than getting a feel for the terrain and vegetation on the mountain closest to camp. As expected, the climbing is not a lot of fun. I guided in this bay years ago and have experience in this regard. The west side of Uyak Bay has always been bad in terms of brush and very difficult to navigate. The east side aspects are easier to move through; but it's a boat ride away!

During our little hike, we see both deer and bear sign. Most of the deer tracks and scat are of movement around the mountain at particular elevations. Most of the movement seems to be on the first benches off the beach. We climb to as far as the second bench, about 750 feet above the high tide line. Even from above, the brush is thick enough that glassing is problematic, with few openings. Anticipating oncoming dusk, we begin our way back down.

We find a marshy little opening and take a break to glass a bit. Voila! Two deer appear. At first they appear to be a doe and fawn, but more careful glassing shows the larger animal to be a spike buck. We discuss the fact that we are very close to the cannery and little packing will be involved. Jason should probably shoot this buck. He appears to be a fine meat deer. Jas is not quick enough on the draw and both deer fade into the brush. A moment later they walk back out. While they went in with the spike in the lead, the smaller deer comes out first and Jason lowers the boom on him; he is a button buck that will dress around 60 pounds. Perfect meat deer! We will divide him in half and be back at the meat room at the cannery in short order.

Except that the spike buck now steps out at under 50 yards, curious as to where his buddy went. Jason wants to fill his freezer, so I bring the old muzzleloader to bear on the second buck. When the smoke clears, we now

Kodiak: One Shot Per Day

View of Larsen Bay cannery from out in the bay closer to Amook Island.

have twice as much meat to pack down to the beach. Now we have to act fast to field dress both deer and head off the mountain. Darkness is coming on quickly; poking around in the dark, in thick brush, smelling like deer blood is not advisable due to the proximity of those pesky and abundant brown bears! Both deer are rapidly deprived of their vital organs. The feet and head are removed from the smaller one and the remaining portion is lashed to my packboard. Jason has brought only a day pack ("Gee Dad, I didn't think we would actually shoot a deer this afternoon....!") With a helping boost from me, the whole spike buck is hoisted to Jason's broad shoulders. The spike weighs about 90 pounds dressed, but is no match for Jason's 285 pounds of "spring steel and sex appeal." We are soon back at the cannery with both our deer hung safely in the meat room.

After some cool drinks and a brief rest, we skin our deer to help them cool thoroughly. In the course of the meat trimming, we get to meet the camp

Kodiak: One Shot Per Day

mascot. Nosey is a gigantic old cross fox who has lived at the cannery for the seven years that Steele has been the caretaker. That is quite old for a wild fox. Apparently the scrounging of scraps has worked out well for Nosey. He is one fat fox.

Steele comes back to the cannery with the skiff just at dark. He has been over at Amook and reports those fellows have a deer already as well. Jason and I wash up and I disassemble and clean my muzzleloader. Once the task is complete, I make dinner. We brought a couple of beef steaks. Being an experienced hunter, I decided it was best not to assume we would have fresh venison available for supper the first evening! I fix those steaks, along with noodles Alfredo and tapioca pudding for dessert. Life is good.

Steele finishes his own supper and comes over at about 9 pm. We talk for a bit. He guided in GMU 19 for a while back in the 90s. We have a number of mutual friends; Steele at one time worked for Stan Frost at Farewell Lake. After a while Steele retires to his cabin behind the main cannery. Jason steps out for a final "drain off." Back inside, he reports he felt "something" on his shoe. He looked down to see an ermine perched ON his foot! We turn in with the sound of light but steady rain on the metal roof. During the night, both the rain and the wind pick up in intensity.

The next morning features a temperature of 40 degrees, pounding rain and a 25 mph wind out of the southeast. There is heavy wave action at the beach out our front door. We heard those waves picking up all night. We are anticipating an impressive 22 foot high tide today in early afternoon. I get up at 6 am and putter around the room. Jason gets up at 6:45. Outside, it's still as dark as the inside of a cow. Steele brings over the necessary official "Transporter" paperwork for us to complete.

After daylight, Jas and I walk the beach to the north for about a mile, watching ducks and whales in the bay. After a while we make our way through the brush and climb to the first bench. Then we ease back south toward the cannery. We never see any deer at our elevation, but from the

Kodiak: One Shot Per Day

occasional clearing, we can see deer over on the far side of the bay. Most of the activity is occurring just below the snowline. They are exhibiting some rut behavior. As the activity increases, we should be seeing more deer.

We make it back to the cannery by 12:30 pm. I make some soup to wash down one of my favorite hunting lunches: sardines and pilot bread. This is the exact sort of lunch that makes you run up mountains, shoot deer, fight off bears and pack out meat! But instead of embarking on this advanced level of adventure we defer to the weatherman. Since the weather is crappy, Jason takes a nap and I go walkabout for a look-see around the cannery. In its heyday, this was a Whitney-Fidalgo operation. It must have been a huge volume operation when it was up and running. It closed in 1983, but obviously operated for a long time prior to that. There are thousands of square feet of warehousing, sleeping quarters, dining hall, store, offices, dispensary and wash building. Those facilities are in addition to the processing facilities themselves. There are also three "over-seer" cabins and three huge generators that powered it all. Now the structures and equipment are rotting and crumbling away. Little or no maintenance has taken place lately. I inquire about a squishy portion of the main dock. Steele says some fuel was delivered and the wheels of the forklift he was using to move the drums punched through the decking of the dock. Yikes!

At 2 pm, it is still raining at the cannery. A thousand feet further up the mountain, it is snowing. We can see only about a third of the way to the top of the peaks on the surrounding mountains. Steele has gone by boat across the bay to Amook to cut and split firewood for the other cabin. We make a 3 pm radio check and assure him that we are smart enough to stay indoors out of the weather. With a couple of deer already hanging in the meat room, this is not a terribly difficult decision. No pressure at all. Jason tries a bit of fishing from the dock, but has no luck.

Steele returns from Amook; he is thoroughly drenched. Without much additional effort or a drop in temperature, he could be classified as mildly hypothermic. He dries off and comes over to chat for a bit. At dark he starts the generator. I grill venison steaks, and add Cajun rice/beans, canned peas and fruit salad to the supper menu. It is now snowing lightly at sea level, but

Kodiak: One Shot Per Day

ambient temperature hovers around 35 degrees. The snow does not stick; it melts as it reaches the ground.

I hit the sack at 8:45 pm and am happy to do so. Steele comes over and invites us to come watch a movie at his cabin. Are we "roughing it" here, or what? I decline the invitation, but Jason takes him up on his hospitality. They also discuss some plans for tomorrow's hunt. Jas tiptoes back in around 11 pm and crawls into his bunk.

I awake the next morning at 5 am, but lay in the bed and read for an hour waiting for Jason to awaken. Then I light the propane lamps and cobble together some breakfast. Steele does his radio check with the hunters at Amook and then comes to see us. For a change in routine, we will "road hunt" today from the Shelbee D. This is a new type of adventure for me. I've never "road hunted" from a boat before. It entails cruising the beaches in said boat, looking for deer. Apparently it is common here, but I have no experience in this regard. Steele, as a licensed "Transporter" cannot do anything but provide "transport" in the field. Any spotting of our quarry is by law left up to Jason and I. Sounds like as good a plan as any to me. I'm told it's a sure-fire way to collect venison, but not trophy bucks.

We make our first run across the bay to Amook and proceed down its west beach. Then we head south up Uyak Bay about 10 miles toward the head of the bay. At one point I spot a couple of fairly nice bucks on the beach. Jason and I get put ashore to take advantage of the "sure thing." By the time we ease into the bight where the bucks had been, they are nowhere to be seen. Turns out it is true; there is no such thing as a "sure thing!" We motor back north along the west side of Uyak. After a bit we spot two more bucks. Jason shoots one of them. We head back to the cannery to skin and cool Jason's buck and to have lunch.

At 2:30 pm we head back out on the Shelbee D. The plan is for Steele to drop us on the beach south of the cannery where a stream, intercepted by a series of beaver dams, flows into the bay. I point Jason toward the most promising looking area for deer, while I circle around to the southwest and make my way to a viewing spot a bit upstream. In the course of getting there,

Kodiak: One Shot Per Day

I see only a lone doe. Unbeknownst to me, at some point I also lose the ramrod out of my muzzleloader. It must have come loose in the thimbles. As I pushed through brush, it seems to have slipped unseen all the way out. Oops. Now not only do I have a rifle that only shoots once before being reloaded, I have a muzzleloader that will only shoot once until I find another ramrod! There is no sporting goods store here in the wild, wet, wilderness! For this afternoon's hunt, I will have just one shot! I hope I can make that one shot count. With luck, the one shot will not need to be aimed in the general direction of a hungry brown bear! I plunk myself down against the base of a large cottonwood tree and ponder my situation. Truth be told, it might be safer if NOTHING shows up and walks by!

But, as luck would happen, after about three hours of sitting I see a pretty fair Sitka black-tail buck making his way in my direction down a deer/bear trail, parallel to the stream course. The buck seems to be headed toward Jason. Although he passes my position at just over 20 yards, I let him go and "will" him toward Jas. I am afforded multiple chances to kill this buck, but I would prefer Jason have the opportunity. The buck crosses a little hogback and heads right for Jas. When Jason catches movement of the oncoming deer and turns his head to get a better look, his movement spooks the buck who then races up the hill... right past me.

It is suddenly apparent if anyone is going to kill this deer, it is going to have to be me. The buck stops in the clear, 45 yards above me and cranks his head over his shoulder to look at me. I have him in my sights and squeeze the trigger of my muzzle-loading rifle. The buck cartwheels and drops. Jason comes up behind me upon hearing me shoot. I direct him to where I think I saw the buck tip over. "You go look son. Your rifle shoots multiple times...!" The buck is dead right there. I explain to Jas that if we are set upon by a hungry brown bear, it's "all on him...." I on the other hand, with no ramrod and thus no ability to reload, will be reduced to beating said bear to death with the butt of my now completely empty muzzle-loading rifle. At least I have a 285 pound stout young lad with a centerfire rifle to protect me!

The buck is a beautiful three-by-three with eye-guards, totaling eight scoreable points. His body size and weight are impressive. We drag him

Kodiak: One Shot Per Day

Jason with spike buck.

into a clearing, take some photos and then remove the innards. Jason starts dragging the deer toward the beach. He doesn't get far before realizing that dragging the behemoth is not going to be practical. It's not practical mostly because of the size of the deer. It's also not practical because dragging a deer in this part of the world is akin to chumming for brown bears. Not to mention the plethora of beaver-downed cottonwood trees that we have to navigate under, over and around. I cut the deer in half and tie both hind-quarters, still attached via the pelvis, onto my pack frame. Jason simply lifts the remainder of the deer onto his broad shoulders. We start off again toward the beach. We reach the beach in good time; Steele is waiting just off-shore with the Shelbee D. We are loaded and well on our way in short order. More to the point, we are safely on the boat before any brown bears catch up to us.

Kodiak: One Shot Per Day

Darkness falls during our trip back to the cannery. In the meat room, Jason and I skin and quarter the buck. I cook up supper for all three of us. Backstrap, onions, rice and green beans. Steele contributes a couple of beers from his store room for himself and Jas. It is cooler now; 30 degrees. Steele advises leaving the gravity fed water line in the sink running tonight to preclude a freeze-up. We hit the hay.

Morning breaks clear (for a change!) and 27 degrees. The water in the sink was still running when we got up, so either the tactic worked or it wasn't all that cold last night. I'm up for good at 6 am. Quick breakfast, then off on the Shelbee D at 7:15. We motor across the bay on fairly calm seas, pick up the other hunters haul them to an open point a bit further up into Uyak. They will be assaulting the mountain today. We cruise on further up into the bay. By 9:30 am we have seen a nice brown bear and nearly two dozen goats. For some reason there are fewer deer on the beaches this morning. We go as far up the bay as the tide will allow, then turn and head back north. We spot more goats and finally some rather nice bucks, but those deer are high on the mountainside. The other guys should do just fine today.

On Steele's fish-finder we get to have a look at the wreck of the *Aleutian*. This was a large vessel that sank in Uyak Bay May 26, 1929. The location of the wreck remained a mystery to most of the world until just a few years ago. Steele knew where it was, but did not at first realize the form appearing on his fish finder was a shipwreck a lot of people were trying to find. Finally someone asked the "local expert" and the mystery was solved!

We end up back at the cannery for a late lunch. After some consultation, we ask Steele to call Seahawk Air on his sat phone. With some bum weather on its way, it would probably be wise to be picked up tomorrow morning, instead of waiting another day and taking a chance of not getting back into Kodiak in time to make my flight that evening. I need to be in Anchorage that night in order to make the Board of Game meeting the following morning. Steele calls and determines that while Seahawk cannot come get us tomorrow, they CAN come tonight. We hasten to pack up and make ready for Rolan to show up with the Beaver.

Kodiak: One Shot Per Day

This was the nice Sitka Blacktail buck that luckily fell to one shot. The only shot that my muzzleloader was going to make that day.

We race back to the meat room. The reason Steele was so adamant about shutting the door completely is now clear. Three or four magpies found their way into the meat room and are helping themselves to bits of venison. Jason gets to flailing at them with my sharpening steel and manages to whack me a good lick on my right hand. I buck up, the birds fly out and we work feverishly to trim and pack meat into tubs and wet-locks for the flight.

Rolan lands about 4:30 pm in the Beaver. Soon we are bound for Kodiak with all our gear and 150 pounds of boned venison. With clear weather, our return route is more direct than the one we experienced on the trip over from town. There is more snow on the mountains now. We see a lot of goats; most are right at the snowline.

Kodiak: One Shot Per Day

Back at the float dock in Kodiak, I settle up with SeaHawk Air. Then Jas and I drive back to his house. Sue is soon home from work. She and Jaden are glad to see us and probably darned grateful we showered before the reunion.

It's overcast the next morning in the City of Kodiak. Sue goes off to work and Jas and I have breakfast at King's Diner. We take a few minutes to visit the ADFG Area Biologist at his office. Then we take a nice drive out toward Chiniak and Bunker Hill, south of town. Lots of fun things to see, including Jason's new favorite duck hunting spots at Deadmon Bay, Women's Bay, Middle Bay, and Kalsin Bay. We see a total of 13 different species of ducks!

Finally we head home, stopping in town to buy some crab for a feast tonight. We finish prepping the deer meat. I'll take about half a deer. Jason drops the remainder of the venison at the processor. The crab feed goes exceptionally well!

The next day we visit Mack's Sport Shop where Jason and Jaden are "regulars." He drops Jaden with a cooing sales lady and we tour the store. I haven't been in here for years, but it is one of my favorites and little has changed with the passage of time. On to the post office where I mail some gear so I can stuff my duffle with frozen venison for the flights home. I also mail Steele his FRS radio I just found in my pack. In a few hours I am at the airport and bound for Anchorage for an Alaska Board of Game meeting.

It's been a nice trip; one I have dreamed of for nearly 27 years... when my son would take ME hunting (as long as I paid for the charter flight!)

Chapter 20

The Big Chill

September 1992 was an interesting month when it came to weather. The weatherman ushered in some of the coldest September temperatures ever recorded across Alaska. Global Warming was but a fleeting memory; we experienced something more akin to a brand new Ice Age. I continued on with plans to hunt some clients on the Sixtymile River, north of Bettles on the south side of the Brooks Range. It was a mixed bag hunt and we had a couple of New Mexico hunters.

It does not make sense to shuttle all the gear, staff and clients from Fairbanks in a small bush aircraft, so for this hunt, Bettles will be the point of outfitting. We start by sending guide Sharon McLeod-Everette, packer Bill Mark Miller and hunter Wendel of Espanola, NM on the afternoon Frontier Flying Service flight to Bettles. They will overnight in Bettles and be ready to fly to camp tomorrow morning. I finish the grocery shopping and take a load of freight over to Frontier for a later freight run to Bettles.

The next morning I am up early to begin loading a Super Cub at Tamarack Air on the East Ramp at Fairbanks International Airport. Pilot Jack and I are all ready for a 7:30 am departure when along comes the FAA inspector demanding we take a time out to do a "ramp check." The only thing he finds to gig us on is that the inspection date is worn off the aluminum inspection plate on the fire extinguisher. All the important "weight and balance" issues are fine and all paperwork is in order. We quickly replace the fire extinguisher and Jack takes off with a full load of freight and gear to go directly to the strip at camp. I pick up clients Mark and Sam at their hotel and haul them to Frontier. We leave Fairbanks on time at 12:15 pm and make it to Bettles non-stop in just over an hour. Jack has just come back in from camp and is refueling the Cub. Perfect timing! I send Sharon to camp on the next load. This gives me time to talk with the clients and Bill. I take the next trip in. I see a dozen moose along the John River, but none in the Sixtymile. Jack reports that he saw two black bears within five miles of camp.

The Big Chill

Water dripping from a canyon wall quickly became icicles. Lots of them!

Once we are all in camp, we sight in the hunters' rifles. Both are fine. I glass for a bit, spotting a black bear just down from camp on the opposite side of the river. There are 20 sheep on the mountain above camp. Wendel spots a wolf across the river. We have a fine supper, a nice campfire and sack out by 10 pm. Tomorrow we hunt; then all will be right with the world!

It is cool the next morning. Bill and I are up early while the the hunters stay in their warm sleeping bags until after 7 am. The ice on the water bucket is not a good omen. It's only September 7, although this IS the Brooks Range. Sam is a non-hunting observer. He decides to stay in camp. We already knew he could not walk well and probably could not keep up with our normal hunting pace. However, Sam has also announced that he is not willing to stay in camp overnight by himself. This means if the rest of us need to spike out in Organ Creek this year, I will need to leave a packer or guide just to be with Sam at main camp. This is not a deal-breaker, but it is a logistical consideration.

The Big Chill

After breakfast, the entire gaggle (less Sam...) heads downstream and across the river toward the pass over into the Malamute Fork of the John River. We are seeing lots of wolf tracks. We spot a nice chocolate color phase black bear, but Mr. Bear is safe for a bit. We are headed into sheep country to try and find a ram for Wendel. In an attempt to liven things up, both Wendel and Bill take impressive back flips into the cold creek. Sharon is wet too, but that is the result of improvised hip boots. She has tied a black plastic garbage bag over each leg and then slipped her boots back on over them. Only Mark and I remain fairly dry this morning.

We spend the better part of the day combing the drainage for rams, but do not spot any. On our way down, we do see the chocolate-colored cinnamon bear again. Mark, Bill and I take off to stalk the bear. Sharon and Wendel will make their way back down the drainage and then upriver back to camp. To reach the bear, we have to traverse a mile of the mountain we are on, dip down through the creek and then assault another 3,000 foot "hill" through fairly thick alders and willow brush. It takes a full two hours to get above the timber on Mountain #2. We make a fruitless search of that mountain top... no bear. Not one to give up that easily, I make a command decision to make another sweep. Perhaps we missed him in a dip or behind some brush? This time we find the bear. He is upwind and below us in elevation, a perfect situation. Mark and I stalk to within 30 yards of the bear and Mark takes him cleanly. We take some photos, skin and butcher the animal. We are on the trail heading out by 6 pm and back in camp two hours later with a very nice bear hide and some tasty blue-berry infused bear meat!

Supper features excellent steak sandwiches followed by a stout bread pudding made in the Bake-Packer oven. Sharon cooks while I rough skin the head and paws of Mark's bear. I'm grateful to Sharon for her chef duty. I'm also anxious to hit the hay. It was a long day, made longer when I hyper-extended my left knee coming off the mountain in tussocky terrain, while packing out Mark's bear. I can't put a lot of weight on it when it is out straight. A hot bath would be nice, but in this environment, that ain't happening any time soon!

September 8 dawns with a high overcast and not terribly cold. I decide to make it an easy day. We will hunt the creek right above camp. It's a long walk

The Big Chill

Our collection of hunters starting out for a day in sheep country.

up the bottom to get beyond the timber, but there appears to be a "sheepy" looking bowl at the head. The first impediments are three waterfalls that look very nice, but are a problem to circumnavigate. We finally get around the waterfalls and make our way up to about 4,200 feet elevation. On the way up, I spot the very tip of a sheep horn sticking out of the gravel in the creek bed. Probably winter kill or wolf fodder from some previous year. It is quite worn and weathered, but I mark the spot to pick it up on the way back down. We find another post mortem sheep a while later where the main drainage splits. This is quite strange. It is the entire carcass of a young 7/8 curl ram. It is a fresh kill, but with no readily apparent cause of death. No bullet marks, no tooth marks. Nothing. One front leg is gone, severed at the top of the humerus. Some ribs are chewed through (wolverine perhaps)? But otherwise, no obvious trauma. A wilderness mystery that taxes our forensic skills.

The Big Chill

Now we are high enough that we are starting to see sheep! Eight rams in one bunch and 11 in another. Now we are getting somewhere! One is a heavy, obviously older, ram with broken horns and very heavy bases. The right side horn is broken all the way back at half curl. We watch them a while. I would shoot that old ram in a New York minute, but Wendel opts to pass him up and look for a ram with longer and more intact horns. It is starting to snow and visibility is going all to hell. So much for a short, easy day! Despite the late hour, I stop on the way back and take the time to dig out and rescue the old ram horn I spotted on the way up this morning. We don't get back to camp until 7:30 pm. Sam stayed in camp again today, but he had a serious culinary project to keep him out of mischief. He has fried up some bear tenderloin in bacon grease, whipped up a killer chili sauce and cooked pinto beans with chicos (dried sweet corn) in them. It is, simply, New Mexican fare that is to die for! Before turning in, I put a tape on the ram horn I picked up today. That rascal turns out to be 43 ½" in length! That is a damned fine ram anywhere and particularly in this area on the south side of the Brooks Range.

The next morning is cold and quite snowy. We sleep in until 7:30. It has snowed, off and on, all night. It's now "crisp" and overcast, with a stiff wind from higher in the Brooks Range. We are going to again try and make it an "easy" day. Sam volunteers to stay by the stove in the cook tent!

Our little band of hunters heads upriver this time. From the mouth of Wetzel Creek we spot a large grizzly in the second canyon on the right, heading up river. We move to the mouth of that side canyon, but now cannot see the bear. The cold wind is murderous, plus most of us again have wet feet from the numerous creek crossings. I wore Xtra Tuff rubber boots today instead of leather, the price is that I end up having to ferry (via piggyback) all the others at every river crossing!

At the mouth of the canyon that contained the bear, we build a little lunch fire for soup and sausage. Afterward Sharon heads back with Wendel. I take Mark and Bill and press on. We go up to the fork in the creek, but see no fresh sign of any game. Our grizzly has pulled a disappearing act and done a damned good job of it! We find a ribbon that Tinker tied last year, marking a place where a slide has clogged the creek with mud and blowdown. We go

The Big Chill

around it in the brush. By 4:30 pm, we have had enough brush busting and turn back. We are in camp by 6:30 pm. Sam is all twitter-pated over his wildlife sighting today. He had asked previously about whether there were caribou in the area. I had 'llowed as how there "weren't any." Midday, Sam is looking out the tent flap and spies some creature headed downriver. As Sam tells it: "I theeenk I see a greezzly...." When the animal gets closer, it turns out to be a wounded cow caribou, probably crippled by wolves or a bear. Sam is left on high alert for predators!

Most of us take turns bathing and washing our hair in the wash bucket tonight. We also work overtime to try and dry out our boots. Dinner is battered halibut with Cajun spice as well as green beans and a zippy chili sauce. Despite some serious exercise no one in the camp is losing weight on this hunt. Sharon graciously finds the time and inclination to sew up a couple of rips in the wall tent that serves as our kitchen and living area. Since the "sewing stuff" is out, she also sews up some serious "air conditioning" in Bill's dungarees.

Morning brings clear skies, but temps are impressive. It is nine degrees BELOW zero this morning. My southwestern U.S. clients were not aware that it got that cold in September, even on the Weather Channel! Truth be told, it is NOT common to be this chilly at this time of year on the south side of the Brooks Range. Mark's tendonitis is bothering him again this morning. His ankles are very painful and wearing boots makes them feel worse. He opts to remain in camp. I urge him to elevate the feet to reduce the swelling and inflammation. He might even ice them if he can do it without freezing his toes!

Sharon, Bill and I take Wendel and hunt downriver toward the mineral lick. Several miles below camp I spot what appears to be a kill of some sort on the river bank ahead of us. After some careful glassing, the hair appears to be that of a grizzly. I approach very carefully and find that it is indeed a dead grizzly. The bear is as dead as Russian capitalism. It appears to have been dead for several months. The hair at the kill site is long, winter hair and the front claws are long and unbroken. Grizzlies don't wear down their claws by digging while they are in their winter dens. Another clue is the numerous

The Big Chill

maggot hulls on the carcass. This bear probably died back in the spring soon after it emerged from hibernation. My best guess is that it was killed by another grizzly in some sort of territorial dispute, although it is obviously a large, older, male grizzly. I scoop up the skull and lower jaw. We find all of the front claws and some of the rear ones. Since "Rank Hath Privileges," I claim the skull, baculum and three nice front claws. Bill gets a front claw and three back ones. Sharon and Wendel split the remaining front claws. Wendel roots around in the maggot hulls and finds a back claw I missed. He says he plans to keep it for good luck until he shoots "his own damn grizzly bear!"

We observe no game near the lick, so we head back upstream toward camp. Halfway back we spot two nice rams high on a bench in the canyon a mile downstream from camp. Sharon takes the spare gear. Bill and I take Wendel and head up after the sheep. It's a long hard pull through the timber. Just as we break out onto a steep rockslide, the wind changes and it begins to snow. Sharon is carefully watching the rams in her binoculars. She sees where they head, but there is no way to relay that information to us. We go ahead and climb the bench and carefully search for the rams, but to no avail. We decide to get the heck off the mountain since visibility has dropped to under 100 yards and we don't especially want to die young.

We leave the bench at 6 pm in heavy snow, making it back to camp in just under two hours. The only game we spot on our stroll back to camp is three cheeky spruce grouse on the main trail. Wish we had a .22! We are getting a little thin on meat. In fact, we have talked about going back to the bear carcass for a few more pounds of meat, perhaps a ham?

Earlier in the day, Mark had boiled out his bear skull. Sam announces that he watched Mark eat the head meat and tongue for lunch. Says Sam: "It wasn't even medium well done. It was a leetle bit dreepy!" But bless their hearts, Mark and Sam have also cooked up beans, with chicos, fried pork and chili sauce. Excellent fare and much appreciated.

The next morning it is still very cold. It is still snowing and there is plenty of snow on the ground. The ceiling is low and the visibility under it is nil. We arrive at consensus that it is probably a pretty good choice to stay in

The Big Chill

camp, at least for now. I need to finish working on Mark's bear hide and do some other chores. Wendel is obviously feeling the effects of yesterday's unsuccessful assault on the Dall rams and Mark's Achilles tendons are still bothering him. I'm fleshing on Mark's bear in the cook tent when Mark comes rushing in from out by the river. He has spotted a large bull moose on the mountainside directly across the river!

With the snow and limited visibility, it's a "now you see him; now you don't" situation. I ferry Wendel, Mark (in sneakers!) and Bill across the river on my back, one at a time....! Sharon and Sam watch from camp, ready to pitch in and give us hand signals if necessary. The heavy snow and swirling wind do not bode well for this stalk, but in my opinion, Mark's feet are bad enough he should take advantage of this gift from the gods of the hunt of a nice moose, this close. It's not every day you spot a trophy bull moose right from camp. Where we last saw the bull is not really a bad pack to the airstrip. It's an all downhill pack to the strip too. Not to mention the fact that we sure could use some moose meat!

Bill and Wendel pull off part way up the mountain and find a perch out of the wind to watch the show. The bull is rudely moving further up the mountain. Mark and I are finally within range. Mark takes a shot at about 70 yards, but for some reason does not connect. The bull begins hastily moving off. I let out a hearty grunt. Mr. Bullwinkle stops, turns and stares at us. This time Mark hits him just fine; the bull is down for the count and never gets back up.

But now look at me. I was fleshing a bear hide and just grabbed my rifle and ran out of the tent. I have no pack. Just one rather dull knife and no sharpening steel! Mark ain't much help either. He has only a pocket knife! After a bit, Bill and Wendel make it up to the kill site. Mark and I get the moose about 2/3 skinned and one front shoulder peeled off.

We slip the front quarter into Bill's pack and he heads off down the mountain to camp to fetch a packboard, knives and a steel. A few minutes after Bill leaves, Sharon shows up with drinking water, snacks, a steel and two packframes. Bill gets the first quarter to the air strip on the gravel bar, then comes

The Big Chill

This is Mark's moose that we spotted from camp. Once it was down, it was all downhill to the airstrip.

back up with another external frame pack. By this time we have the meat all subdivided and well under control. Four loads go off the mountain this time. Bill and I each take a hindquarter. Mark carries miscellaneous gear and rifles and Wendel packs the antlers. Bill and I go back up for the last round. We come off the mountain with way too much weight in our packs. Bill takes another front quarter and a side of ribs. I pack the cape, the boned neck and other side of ribs. Down at the river we build a crib of drift wood and put the meat on it. From camp we can easily see if a grizzly bear attempts to take unauthorized possession of the meat. We cover it all with a blue tarp and retire to the cook tent. This was quite an adjustment to our "easy" stay-in-camp day! It's now supper time; Sharon has a big pot of spaghetti ready. Wendel makes an announcement (now that Mark has two animals, and Wendel has none.) "Today I finally got some blood on my hands... but it was my own!" He has cut himself while helping to skin Mark's bull!

The Big Chill

The next day begins overcast, cold and still spitting snow. We are going to try one more time to have an "easy day." Mark's tendonitis was not helped at all by running up the mountain and shooting a moose, even in sneakers. I finish the bear hide and flesh the moose cape, turn the ears and split the lips. Once both the trophies are salted, I can finally relax a bit. Not that the capes would decay very rapidly in this weather, but they will now last for weeks! I also make some field expedient repairs to my pack-frame. One strap and several grommets gave up the ghost on the last load of moose meat. Bill stays warm by chopping wood and digging a new outhouse hole.

This afternoon, I take all three guests down below the airstrip to fish for grayling. Five nice fish will be joining us for dinner. We fetch them back. At 2 pm I set up a rack of moose ribs on the fire pit. By 8 pm they smell so tasty we can hardly keep our lips off of them. I've mixed up some rice, corn and tomatoes and made cornbread in the Bakepacker. After the first batch of cooked moose meat is carved off the rack, I set the rack back on the fire. The quote of the day is from Sharon, who gazes out the flap of the cook tent a few minutes later and innocently asks: "Are those moose ribs supposed to be on fire?"

We are now on Day Seven of this chilly adventure. It is cold, but at least the skies are clearing. Mark's heels are not feeling much better, so I take Wendel and Bill and head up Camp Creek to look for sheep. Wendel has now developed a "thing" for the broken horned ram he turned down last week! What a difference a few cold, sheepless days make. We only go a mile or so when I take a bad tumble on the ice. I'm fairly certain I injured some ribs on the right side of my back. I lay pretty still in place for a bit and try to evaluate my situation. Once convinced that there is no bleeding, I pop some ibuprofen and keep going. By noon I am seriously considering going back to camp because of the pain. We stop for a long lunch at the bottom of the chute where we saw the rams previously. As we are getting ready to leave, we spot a grizzly a half mile above us. We quickly shove our gear back into our packs. With help from Bill, I clamber upright again. We begin our assent of the steep mountainside. We need to gain both elevation and a wind advantage to put an effective sneak on this bear.

The Big Chill

Thirty minutes later we are level with the bear. We have at best, a cross wind. But we have the bear in sight again. I estimate the range to be all of 250 yards. Ordinarily I would consider this to be "too far" for an ethical shot at a grizzly. But Wendel is a New Mexico antelope hunter, familiar with long-range shooting. It is open country and there is no brush for the bear to disappear into for 400 yards in any direction. Unfortunately Wendel takes issue with my estimate of the range. He demonstrates this by shooting over the bear three times! Then he cannot find his spare cartridges. By now, the bear is steadily headed for the top of the mountain, although it is not moving terribly fast. A fall-fat bear moving up a steep mountainside has a disadvantage that works to our advantage. I hand Wendel my rifle. I would bet a lot of money he is not going to hit this bear. The grizzly is as good as gone. I won't need my rifle for back-up; Wendel is welcome to use it! Much to my surprise, Wendel rolls the bear on his second shot with my rifle.

We watch the bear for a full five minutes and detect no movement. It takes another few minutes to get above the downed bear and walk down-slope to it. I never approach "dead" bears from below if I can help it. It's "my policy!" As we near the spot, it is clear the beast is, as Hemingway would say, well and truly dead. We take pictures and skin the bear. It's a sow, but quite old and with a large skull. Wendel's second shot with my rifle entered the lower back and ranged up into the chest cavity for a quick kill.

We are back in camp in two hours. Sharon had the day off, but has worked hard anyway. She inventoried our supplies and is now studying for a test of some sort. Sam is involved in making a moose meat carne with chili sauce for supper. Recipe: Heat ¼ cup of bacon grease in skillet. Add ¼ cup flour and brown it. Add ¼ cup chili powder (Sam has brought along his own dried chilis and ground them up on site). Mix well, then add 1 ½ cup of water. Stir in 4 cloves of garlic, finely chopped. Add more water as needed. Bring to a boil and simmer. The longer you boil, the spicier it gets! Sam knows a thing or two about spicy New Mexico food.

Quote of the Day, from Sam: "Watch out for bears Beel, you sleep closest to the meat pole!"

The Big Chill

The next day it is snowing again and very cold. An all 'round miserable day, especially for people from the desert Southwest. Sharon and Bill hunt with Mark up to the mouth of Organ Creek to look for a grizzly. Mark's heels seem a bit better. Today he is wearing his sneakers again... and all the rest of the clothes he brought! He is quick to remind me I had him leave lots of extra clothes in Fairbanks to save on weight when we flew in. Heck, I had no idea we would all be borderline hypothermic for the entire hunt. Where is Global Warming when you really need it?

Wendel, Sam and I stay in camp. My back and ribs are really bugging me. I didn't get much sleep last night. The current situation is made worse due to the fact my Therm-a-rest pad has pin holes in it somewhere and it is losing air quite constantly. The only slightly comfortable position I can get in is lying on my back. Trying to lie on either side is so painful I cannot sleep at all. At 4:30 am I gave up, got dressed and went into the cook tent and built a fire in the stove. After breakfast I finish fleshing Wendel's grizzly hide and sharpen all my knives. I drag in a bit of firewood, but generally my injury is keeping me slow and fairly non-productive.

The hunters are back by 4 pm; the weather has turned even colder. The snow picks up and so does the wind. They've seen no game; visibility is dropping rapidly. Sharon cobbles together another great spaghetti sauce. Mark fries some moose steaks for a "snack." Quote of the Day from Sam: "Sharon, if we were married, I would weigh 300 pounds!" This is saying a lot, coming from a man who barely tips the scales at 140 pounds!

I sure hope we get a break in the weather that will allow my partner Dave to fly his Super Cub in and land on the strip. I need help with the guiding; my back is getting more painful. I've fabricated a brace of sorts with a foam pad, some meat bags and a hip belt from a pack frame. It provides some relief by at least keeping things stationary. I plan to sleep in it. The snow is even deeper by the time we hit the hay.

In the morning, the wind is still blowing and it is even chillier. But at least the snow has stopped. Two quotes of the day on a similar subject. Bill says: "It's a bit nippy out this morning!" And Sam proclaims: "Eez a cold wan...!"

The Big Chill

Wendel's "hail-mary" rifle shot earned him this pretty grizzly bear trophy.

In the afternoon, Sharon and Bill hunt both Mark and Wendel downriver. At about 4 pm they spot a large grizzly on the south side. Crossing the river gets interesting. They have no hip boots, so they cross in their socks and put their boots back on when they reach the other side! Ambient air temperature can't be over 10 degrees. They trundle up the mountain and closer to the bear. But at 400 yards, an errant breeze carries their scent to the bear. He winds them and goes out over the top. Sharon is smart enough to know not to try and catch up. The group heads for camp.

Around 7:30 pm, Dave flies in from Fairbanks. He and his hunter Mike (Sam's son) were hunting sheep at my Johnson River camp. Mike killed a nice ram up on the Spur Glacier. Mike had been intending to hunt grizzly with us and his dad, but he got tired of waiting for the snow to lift and has returned to New Mexico. Sam is of course disappointed he didn't get to spend time in camp with Mike; they had planned the trip for an Alaska adventure together. Sam cheerfully admits that even though his son deserted him and "went south for the winter," he had a lot of fun hunting with us.

The Big Chill

Dave brought some fresh groceries, including real sweet corn! He also has some great stories about the number (and quality) of hunters who were on the Johnson River while he and Mike were camped there. He shares the tale of two nimrods who, not knowing that Dave and Mike were watching them with binoculars, killed a sub-legal ram. Dave made the obligatory phone call to the Alaska Wildlife Troopers. Alaska law is clear. Guides have a legal obligation to report any observed game violations.

Wendel wanders in around 8:30 pm and fills us in on Sharon and Mark's stalk trying to get the big grizzly downriver. We have a great dinner, with two entrees. Pork with chili, and fried moose heart. It is getting a lot colder. Even the Sixtymile River is freezing up!

In the morning the mercury has dropped to -1 degrees! Even the thermometer on Dave's Super Cub in full sunlight at 10:30 am registers only 10 degrees. We eat a big breakfast and head down to the strip. It's cold enough the Super Cub won't start on the battery. We haul down some stove pipe and fire up a mountain stove to "hotpot" the airplane's powerplant. After 30 minutes or so, the oil is warm enough that Dave is able to hand prop the airplane and get it running. Dave first takes Sam for a "flightseeing" trip up the Sixtymile. Sharon has a class she needs to get to in Fairbanks, so he flies her to Bettles so she can catch the afternoon Frontier flight to town. Bill and I move moose meat from the meat pole down to the air strip on the river.

On his next trip, Dave takes some of the moose meat to stash in Bettles. I'll have Tamarack Air take it to Fairbanks on their last trip south tomorrow. Dave takes a final load of meat (hindquarters) and the trophies. The meat, hides and cape fly internally. The moose antlers are lashed solidly to the wing struts. Dave will fuel up in Bettles, then fly back to Fairbanks.

On one of his shuttles to Bettles, Dave spoke with Tamarack pilot Jack. Jack originally planned to stay overnight in Bettles and start ferrying us out first thing in the morning, but Tamarack is pretty backed up due to weather issues in the Alaska Range. Apparently it has been even worse there than here in the Brooks Range. Four days of snow in Fairbanks have really stacked things up. Tamarack even changed over to wheel-skis on a couple of planes

The Big Chill

to be able to deal with the snow in the mountains where they need to land to pick up their hunters. There is over a foot of snow in Fairbanks, and even more down in the Alaska Range.

We do some sorting and inventory of gear, then retire to the wood stove in the cook tent to hide from the bitter cold. As soon as the sun goes behind the mountain, the mercury drops back down below zero. The Sixtymile River has frozen all the way across in most places! There is little moving water on the surface. Thanks to Dave and the grocery run, we feast on beef steaks, corn, fried potatoes and peas.

As night falls, I step out to toss some dishwater. I hear footsteps on the shelf ice along the far side of the river. It's another bull moose. I roust everyone out to see the bull. I'm able to call it and get it to stop. Mark then tries out his New Mexico version of a moose call. The bull is not at all enticed. He does a rapid "about face" and hauls ass off into the brush!

Our final morning in camp dawns bright and cold. We begin packing to leave. Bill Lentsch flies in around 10:30 am. After hearing about all the weather problems in Fairbanks, we had not expected him this early, but we rise to the occasion. He doesn't have to wait long before we have him loaded and ready to fly. I explain that I want the three clients to make the 2:30 pm Frontier flight out of Bettles to Fairbanks. Then he can move Bill and I to Bettles, where we will spend the night. Sam is the first out, then Wendel and finally Mark.

Bill and I work feverishly to get the storage drums packed and hauled up into the cache. Bill leaves on the next flight and I'm left to take down the cook tent, smash cans in the garbage and hang the last of the gear. I make a mad dash to the strip just as Bill comes in for the final ferry. I'm in Bettles by 4:45 pm. Bill and I will have to stay overnight, but at least the clients made it onto the afternoon flight to Fairbanks.

Bill and I get beds at the Sourdough Bunkhouse. More importantly, we take long overdue showers! There is a going-away potluck at the Bettles Lodge for someone we don't even know, but we attend and have a great meal! I bump

The Big Chill

Mark's cinnamon bear was an interesting color phase not often seen in the Brooks Range.

into some Bettles friends at the party, including Bill and Lil Fickes and Dave and BJ Schmitz. It's cold and windy as we retire to our cozy places in the bunkhouse. The next morning at 11:30 am, we board the morning flight to town. A great hunt with great people...! But I grant you..... it was chilly.

Note: Subsequent X-rays provide evidence I had cracked three ribs when I fell on the ice during the hunt. No wonder it hurt!

Chapter 21

Pitch 'Til You Win!

This chapter chronicles what might be called a "typical" fall hunt, circa 1993. In those days I was offering mixed species hunts in the Brooks Range. Normally we would hunt caribou and sheep in early August, then move our base camp from Galbraith Lake to Happy Valley to hunt moose, caribou and grizzly the first couple weeks of September. Weather was the deciding factor on timing. We tried to be out of the mountains by September 15 to avoid bad snow.

We leave Fairbanks early Sunday, August 29th. We head up the Haul Road, cross the Yukon River and arrive at Coldfoot shortly after noon. Dave is there with the Super Cub. Hunter Joe will wait with Dave to fly through the Brooks Range, hopefully through Atigun Pass. Hunter Doyle will stay with us and the trucks. Our friend George and his son Phil are moving some drums of aviation fuel for us; in return, Dave will fly them out for a quick caribou hunt before they head back south. The two Millers, Bill Sr. and his son Bill Mark will drive Joe's pickup on north. Bill Sr. makes his famous pronouncement as plans change: "You gotta pitch 'til you win!" This refers to the rut some find themselves in while on the midway at a rodeo or county fair, paying dollar after dollar to get baseballs to chuck at milk bottles or balloons, for a chance to win a $3.00 stuffed toy! On northward from Coldfoot we go. While Dave could have made it to Chandalar Shelf just short of the pass in the aircraft, Atigun Pass itself is marginal or below and thus, not a possibility. Underscoring this is our sighting of a freshly wrecked Cessna 172 lying on a pile of rocks on the north side of the road, near the top of the pass! In Coldfoot we heard the pilot dropped off a friend at Deadhorse and was attempting to get back to Anchorage on a tight schedule. She walked out 300 yards to the road with head and jaw injuries. Pushing the weather when flying in Alaska is seldom a winning strategy. It was not a winner in this case either. She is lucky to be alive.

Pitch 'Til You Win

We stop briefly to check on our cache of drums of aviation gas at a friend's cabin at the Galbraith Lake airport. The gas has been delivered and is fine. Just north of the lake, we encounter dense ground fog that rolled in off the Beaufort Sea. Near Toolik Field Station, a North Slope research center for the University of Alaska, Fairbanks, visibility is down to less than a quarter mile. It breaks up a bit around Slope Mountain and isn't that bad at Happy Valley when we finally pull in around 7:30 pm. Happy Valley is a nice gravel airstrip that was built adjacent to the Haul Road to facilitate pipeline construction 20 years ago. It is used by hunters as a convenient jump off spot for hunting on the Slope. The road trip from Fairbanks to Happy Valley has taken a solid 14 hours. As always, the trip itself is part of the adventure!

We hustle and get a couple wall tents set up. The cook tent includes a small wood stove for heat. The other tent is just for sleeping; it is adorned only with cots. We brought our firewood with us since we are far above the timberline at this latitude. Everyone is in bed by midnight.

It is still cool when we awaken in the morning. The weather features patchy fog and light snow. I must say Doyle was not lying when he proclaimed last night he has a reputation as a "world class snorer." If anything, his proclamation was an understatement! This man can generate some serious noise. George got tired of listening to the crescendo and went to his truck to try and sleep more peacefully. Doyle handed out earplugs when he gave his speech about his snoring. Upon learning that George retreated to the truck, he asks if George had been wearing his. George replies "I had 'em so far in, they were touching each other! And I still couldn't sleep!"

The snow slacks a bit around noon and flying activity at Happy Valley picks up. This is an important hub for pilots in the hunting community. A number of Fairbanks hunters who have been hunting caribou further west on the Anaktuvuk River drop in for coffee and a visit. The passengers have trucks parked here at Happy Valley and will soon head south. The pilots will securely tie down their two airplanes, a Cub and a Maule. They plan to ferry the planes to town later when the weather improves. My guys take spinning rods and head to the nearby Sagavanirktok River (the Sag) to try for grayling. Bless their hearts, they are soon back with enough fish for supper for the

Pitch 'Til You Win

whole bunch. I make a big pot of beans to go with the fish course. We feed 16 people at supper, including the Fairbanks visitors.

Doyle is a custom gun maker back home in Texas. After supper he gets a chance to "admire" my ancient and dilapidated H&R Ultra Rifle in 7mm Rem Mag. It is a working rifle, not a wall-hanger. It has a roughed out, unfinished Chet Brown fiberglass stock. A connoisseur of fine custom guns and fancy walnut stocks, Doyle promptly proclaims my favorite rifle to be "the ugliest damn gun" he's ever laid eyes on! In fact, he says, he "doesn't even want to be in the same tent with it!" And "Now I know why you carry it in the bed of the truck instead of the cab!" Doyle is quite a character! We would love him even more if he would quit snoring.

It is raining lightly, but the ceiling and visibility aren't bad here now. Hopefully things are good on the south side of Atigun Pass. Dave is smart enough to not try to fly through with the Super Cub unless he knows the weather is OK on both sides of the pass and in Atigun Pass itself.

During the night I get up to "drain off." Not surprisingly, I find that the rain has turned to snow. Pretty much our whole world is white. Back in the tent I succumb to a coughing fit. It's so bad it even wakes up our world class snoring champ Doyle. After listening to me for a while, Doyle gets up and brings me a couple of Comtrex to shut me up. Touche, Doyle!

In the morning Don, our expert cook, makes blueberry pancakes for the crew. Luckily Bill Jr. and Phil picked berries last night before the snow buried them all. The snow shows no sign of letting up. In fact it snows until mid-afternoon. Both Bills and Phil head up to the road culvert to try for a fish or two. They have no success, but they are able to avoid hearing more of Doyle's stories about Snuggles his dog, so the time is not wasted. The stories are told every time someone new stops by our camp! The snow storm ratchets up again. My guys are forced to scrounge around for some pallets and dunnage to feed the wood stove. We brought some wood, but clearly we didn't bring enough to outlast the blizzard. The Fairbanks guys come over with some fresh caribou liver and heart. Don fixes it for supper with onions and bacon. Don and his propane oven and range are a VERY popular fixture here at

Pitch' Til You Win

Happy Valley. Everyone downwind stops by for a taste of his wares! As always, dinner is well-received.

The snow slacks off around 6 pm. A Cessna 206 gets in from Umiat and Dave Neel is able to get a couple of caribou hunters flown out to a spike camp.

September 1 breaks cold, with snow spitting intermittently. The ceiling is slightly broken, but not flyable by my standards. We have a leisurely breakfast. Then Doyle asks to "use the phone." This will entail a 50 mile roundtrip back down to the Alyeska Pipeline Pump Station. No worries. We have nothing else to do until Dave gets here with the Super Cub. As we start back from the Pump Station to Happy Valley, the enormity of the real estate situation dawns on Bill Sr., a lifelong Texan. He states it succinctly: "The next time I hear a Texas rancher bragging about his cattle and 5,000 acre ranch," he says, "I'll just let him know that I hunted 25 million acres... without a cross-fence or a cattle-guard. This is truly the Big Pasture!" From now on, Bill Sr. will be known by all as "Big Pasture Willy!"

Along the road, several miles south of camp we spot a guy grabbing his video camera out of the back of a parked camper. We assume (correctly) there is imminent action coming on, so we pull over to watch the show. What we see is a parade. In the front, by 100 yards or so, is a medium sized grizzly bear. Next in the procession is an anxious young bowhunter. Ten yards behind the bowhunter is his rifle-toting buddy. Bringing up the absolute rear is the camera guy. This show is walking away from the road and toward a camp, later found to be that of the hunters involved, on the top of a hill in an old gravel pit. This bruin is what I call a "road bear." He is accustomed to people and traffic along the Haul Road. He seems to be aware of the guys lined out behind him, but is not showing any sign of being particularly concerned about them. He continues on about his daily routine, ignoring the entourage following behind him. In this case, the bear's routine includes a "snack of opportunity" at the hunters' camp! At the camp he finds his reward, an unsecured cooler that is begging to be inventoried by a mischievous bear.

While the bear is batting the cooler around, the hunters close to within 20 yards. When his firearm backup is in place, the archer hauls back and lets an

Pitch' Til You Win

Bill Miller Jr. unceremoniously stuffed into the back of the Super Cub.

arrow fly... right over the bear's shoulder and off into the wild, white yonder. The bear gives the hunter a withering glance. Apparently this is enough to convince young Robin Hood that further aggravation of the bear at this close range might not be as terrific an idea as he originally thought. The hunter confers with his armed wing-man. As the conference ends, the rifle toter sends a warning shot in the bear's general direction. The bear ambles off across the tundra. The hunters wisely start packing up their stuff to move to a different location. We return to our camp with yet another yarn to share! The next morning it is still snowing and blowing. Thanks to sitting around the stove drinking hot drinks last night, I had to get up four different times to pee during the blizzard! The weather lifts a bit by mid-morning. Len fires up his Super Cub and begins moving some of his hunters. A trucker stops in and passes the message letting us know Dave and Joe are holed up at Chandalar Shelf with another guide/pilot. Dave has called back to Fairbanks and is having a friend ship up skis for the Cub! We sit around the stove in the cook tent and inhale the heady fragrance of the muskox stew Don is preparing.

Pitch'Til You Win

Afternoon brings a visitor from Len's camp. He is famed outdoor writer, Aaron Pass, from Georgia. He will be hunting moose with Len. I recall meeting him in Fairbanks years ago when he was the Southern Field Editor for Outdoor Life magazine.

The following day is a lot warmer; it reaches 40 degrees. A Fairbanks dog musher friend, Leroy Shank will be driving south in a little while. He offers to stop at Chandalar Shelf to give Dave a weather report from Happy Valley and through Atigun Pass. Doyle is up earlier than usual; his plan is to pack up his spike camp gear. He figures, and perhaps rightly so, that the warmer weather means we could be getting him flown out soon. But soon he returns from the sleeping tent; the "Bills" are both asleep. Doyle announces "My snoring keeps 'em up at night. I may as well let 'em sleep during the day!" I can now see the tops of the peaks in the Brooks Range to the south. Something should happen today! I celebrate with a bath and a shave. Len flies Aaron out to his moose spike camp. An air taxi guy flies in from Deadhorse and reports dense fog blowing in from the west. Was the break in the weather too good to be true? At 7 pm, Dave and Joe fly in. They made it through the pass; we are ready to get out and hunt.

We hurriedly load Joe and some gear. Dave flies him over to our caribou spike camp on the Toolik River. While Dave is gone I throw my own gear together, along with more provisions and food. We might be "a while!" Dave is back in an hour. I'm the next guy out. Then the Bills and finally Doyle. Dave takes off from the spike camp gravel bar just at dark to return to Happy Valley. We are finally in spike camp and will be able to legally hunt the following morning.

My watch "alarm" is set for 5 am, but it never gets a chance. Doyle is pumped up and ready to shoot something! He gets up at 4 am and makes sure the rest of us are too! We have a good breakfast, then twiddle our thumbs until 6 am when it is light enough to walk safely. As it gets lighter, we can see the river has risen during the night, probably due to snow melt far up drainage in the mountains. Ordinarily this would not be a problem. But we set Doyle's tent up on an adjoining gravel bar is far enough away from the rest of us that we are not subjected to the "mating walrus" sounds of Doyle's snoring. Now

227

Pitch 'Til You Win

Bill Miller and I wait on a gravel bar for Dave to return with the Super Cub to move back to Happy Valley.

the rising waters have reached the back of the tent. We move Doyle's tent to higher ground before leaving camp to hunt for the day.

Soon after we leave camp and reach the river bank, it becomes obvious that there is no way to cross the main channel of the river and remain dry while doing so. Walking on the gravel bars isn't bad, but at the cutbanks of the meanders, we have to go up the bank and fight tundra and tussocks. It is not a simple proposition. Doyle bogs down almost immediately due to knee issues. This hunt could be a long one...

Joe and I head out to some higher terrain on the west bank of the river. From here we can see four moose and about 20 caribou. The caribou are about four miles east, on the wrong side of a cold, fast river. We walk down the hill and back to camp to talk with Doyle and Bill Mark. I find Doyle a nice spot overlooking a tracked-up crossing that eventually is likely to see caribou

Pitch' Til You Win

swimming the river. Bill Mark has been chucking rocks off the strip and had a nap. He's ready for anything.

Joe heads back upriver with me. Against my recommendation, he is wearing leather boots. We find a nice glassing spot and settle in. Soon five bull caribou make a fatal decision to cross the Toolik within a few yards of our observation point. We select the biggest bull and Joe anchors him nicely. As it turns out, an ideal situation would have been to have let the bull get a few more yards away from the river. The mortally hit bull flops back into the chilly water. After pulling the animal out of the water, we skin, butcher and bone it and pack him the two miles back to camp. Dave stops in after flying George and Phil to our other caribou spike camp 10 miles further down the Toolik. He is able to take Joe's caribou antlers and meat back to Happy Valley. He will be flying Don into the upper Sagavanirktok tomorrow for a chance at a Dall sheep. Dave will check on us again in a few days.

We accidentally oversleep in the next morning. When we finally roll out and give the countryside a look-over, there is activity at the natural mineral lick. Doyle proclaims it to be a moose, but with my spotting scope, I can see that it is actually a large black wolf! After a bit of glassing we spot a lighter colored wolf too. Joe announces he is willing to use his moose tag on a wolf. The metal locking tag is legal to use on an animal of "equal or lesser tag value." I'm a huge proponent of that concept! A wolf is a terrific trophy. Plus, that is 1,000 pounds of moose meat I won't have to carry across the tundra and along the river back to the air strip to be flown back to town! Bill and Joe set out upriver to try and get ahead of the wolves. That portion of their plan does not work out particularly well.

I get Doyle set guarding the caribou crossing. Then I head for the glassing hill. From here I can see the two wolves, plus another, at an old kill across the river. In between bouts of glassing, I am able to get Joe's caribou cape fleshed, the ears turned and the lips split. When caribou activity slacks off midday, we all doze in the sunshine, but soon the wind changes and becomes a bit brisk. Doyle would like a windbreak of some sort. I cut some willows and fix him up.

Around 5:30 pm there is another cow and calf moose in sight, plus about 50 caribou scattered within our sphere of influence. The wolves have napped,

Pitch' Til You Win

but are waking up and stretching on a knoll ¾ of a mile across the river. They get up and start to move. Bill Mark gets the hare-brained idea they will be heading to the river for a drink after their meal. He thinks we could ambush them. He takes Joe and sneaks through the willows to a root-wad sticking up out of the gravel bar opposite where it appears the wolves may get to the river. Son of a gun, the wolves show themselves on the far side of the river to drink, at a range of just 40 yards. Joe kills the black one with one shot. I love it when a plan comes together. Even if it's not my plan and I was positive it would not work! One of my most famous sayings is: "If you have a nutty idea, but it works? Then it was not a nutty idea!" It applies here!

Now we have a teensy logistical problem that the master hunt planners have not taken into consideration. The dead wolf is on the "wrong" (opposite) side of the river from us! This looks like it will entail some ingenuity and possibly even some borderline hypothermia! Bill volunteers to try and make the retrieval. We dress him up in all his rain gear and use duct tape to seal off his ankles and wrists. Next we tie some stout cord onto him and launch him out into the Toolik River. He makes it "almost" across before he is summarily swept off his feet. With the string, we reel him in like a salmon. Part of the problem is that Bill is sort of skinny and there is a fair amount of air trapped inside his rain suit. When he gets swept off his feet by the strong current, he tends to float "ass up." It's funny, but not particularly productive in terms of dead wolf recovery! We regroup.

As much as I don't want to endanger a client and perhaps risk what little reputation I still retain as an outfitter, Joe makes a good point. "It's my wolf" says Joe. "I will swim for him…!" With that, Joe strips down to his long johns and makes the chilly swim across the Toolik. Like General Washington crossing the Delaware River on Christmas Eve 1776…. Only without the boat! Once on the "wolf side," Joe ties the line to a convenient anchor on the wolf. This wolf has a radio collar on it. We haul the wolf back to our side. Then we tie a stone on for weight and chuck the cord back over to the far bank. Minutes later, we reel in a soggy, shivering Joe as well. Standing on the gravel bar in soggy long johns, in retrospect, he really doesn't resemble George Washington all that closely. The wolf most closely resembles a German shepherd dog that got caught out in a heavy downpour.

Pitch 'Til You Win

Later we regale Doyle with the story. He has seen caribou from his lookout, but not on the "proper" side of the river. Doyle proclaims: "I don't believe I want any of the caribou I've seen bad enough to swim for 'em!"

It is still cold the next morning. Doyle grabs some snacks and heads for his perch. Joe will stay in camp today in case Dave flies in to check on us. He is out of tags and figures he should head home. The rest of us head out to assist Doyle with glassing for caribou. For some reason, there are few caribou around this morning. Doyle is getting discouraged, but he is cognizant of his physical limitations and maintains a practical outlook. To compound these issues, Doyle in confidence admits to me he is almost out of his medications. His plan is to go on 50% dosage. I'm no physician, nor do I play one on TV, but this sounds like a less than terrific idea to me. We head in for supper; about the same time, Dave flies in. He has time for just one trip tonight. Joe is ready to fly, but I make a command decision. Doyle needs to get back over to main camp at Happy Valley where he can figure out a plan to obtain additional prescription medicine. Doyle packs his gear and crawls into the Super Cub "50Q" for the ride back over to the road and camp.

A heck of a front passed through last night. A lot of wind and rain in profusion. It makes us glad we moved our tents to higher ground early on, when Doyle's tent threatened to become a houseboat. Dave flies in and picks up Joe before noon. A Fairbanks pilot friend at Happy Valley has to fly some moose meat to Fairbanks and has volunteered to pick up any medications Doyle can arrange to procure and fly them back up to us. People helping people in the wild corners of Alaska.

For our next surprise, Dave proclaims that Don D'Amato, the cook, wants to try for a moose. He has his unused non-resident sheep tag in his pocket and there is too much snow to head back up into sheep country. We are now on Version Q of my original Plan A. With this new goal in mind, instead of taking me back to Happy Valley, Dave moves me upstream on the Toolik to a likely looking moose patch he spotted on the way out with Joe. We find a good safe gravel bar to land on and Dave sets the Cub down handily without incident. I start setting up the camp; Dave goes back for the two Bills and Don. By 2 pm we are all on the same gravel bar. The river is very braided

Pitch' Til You Win

This is the wolf Joe took, with tactics I "knew" would not work!

here and with care, can be crossed in hip boots. Even better, there are some "bluffs" about 40 feet high at the river's edge that give us a commanding view of the moosey-looking willow patch. We shinny up the bluff and get to glassing. Time is of the essence; we have only tomorrow to find a moose for Don and get it killed and packed.

We have picked a pretty darned good willow patch. From the top of the bluff we spot five bulls and a bunch of cows. With a good feeling for what the morning should bring, we move back across the river to camp. Dinner is the hot pot of beans and muskox stew Don brought with him from his cook tent kitchen at Happy Valley. Bill Mark goes to the low hill to the east and glasses until dark. He puts the two biggest bulls "to bed" as darkness falls. We will sleep with fingers and toes crossed that the bulls don't move too far overnight. The ever optimistic Bill suggests that if they must move, maybe they could choose to move TOWARD camp and the airstrip for a change!

Pitch 'Til You Win

We awaken at 5 am to high overcast in the almost pitch black. It is chilly on our gravel bar. Perfect moose hunting weather! We grab a quick breakfast and pace impatiently for a hint of dawn so we can safely cross the river and make it to our bluff. The wind is light, but favorable. The breeze is directly from the willow patch to us. Perfect. In less than 30 minutes of glassing we spot a legal bull (50 inches of overall spread.) He looks like the younger one that was with the biggest bull last night. Where is Mr. Big? Some bodacious binocular work sleuths out his hiding place. He is still lying down; just the top of one antler shows. Barely. The sun is behind him, so the antler appears dark instead of light. Noting his location carefully, we cautiously begin our approach and stalk.

All four of us stalk to within 100 yards and drop our packs. Only rifles and cameras from here on. The wind is still favorable. The other bull, bless his heart, has chosen to feed off several yards straight away from us. A convenient dry gully lets us creep stealthily to within 70 yards without alerting either bull. I do not want to chance getting too close. I get Don into a good shooting position. This should be a piece of cake, so I hand my rifle to Bill Mark with the instructions "Back Don up...!" I can imagine what an opportunity like this would have meant to me early in my guiding career. Bill Sr. is poised and ready with his camera. It is "Go Time...!"

I whisper last minute instructions and proceed to grunt loudly like a challenging bull. Nothing happens. Perhaps our bull is deaf? It dawns on me that if there are that many bulls around, perhaps this vocalization will not be effective in getting "our" moose to stand up? I switch calls and try a cow call. Boy, that does it! The big bull jumps to his feet and stares in our direction with just one thing on his mind. As he rises from his bed, two sets of cross-hairs settle on his near shoulder. Don laces him through both lungs with his .270. Bill backs him up with a spine shot from my 7 mm. The bull goes down without moving a step. It's all over in three seconds. The big trophy bull is on the ground before 7:30 am. We spend a few minutes taking photos, then get to caping and cutting up the bull. The honeymoon is over and we are ready to pack close to 1,000 pounds of moose meat to the airstrip for transport from the field.

Pitch 'Til You Win

Don D'Amato and Bill Miller Jr with the terrific moose that they lowered the boom on. Bill was "back up."

As we finish the work, Dave flies over in the Cub. Always trying to help out, Dave spends a couple of minutes looking for a place to land 50Q closer to the kill site. He makes a few passes, then drops into a nearby blueberry patch. Says Dave later: "Well, it looked smooth from the air...!" The big tundra tires bounce over several tussocks and come to halt, with the tail flying too far up into the air for comfort when Dave stands on the brakes bringing the empty Cub to a halt. Methinks that taking off again ain't gonna be pretty either! We are NOT going to be ferrying any meat out of this crappy, marginal landing spot. No way; no how!

Dave walks over and inspects the moose, then returns to the plane and his lumpy improvised landing strip. I can sense he has concerns about taking off. In my opinion, his reservations are well founded. Few options, but lots of reservations. The Bill's go over to the plane with him. They help manhandle the plane around and wrestle it over to the downwind end of the blueberry patch. They spend a good 45 minutes knocking down the worst

Pitch 'Til You Win

of the tussocks and some small willow brush. With Dave ready in the pilot's seat and the prop turning, they await a gust of wind on the nose of the plane. When the wind picks up a little, Dave hauls on the throttle. The Bills run alongside, pushing on the wing struts for as long as they can keep up. With a lurch, the Cub is airborne again. Whew! We are all quite relieved; most of us are nearly about out of "pucker!" Dumping the Cub in a blueberry patch many miles from the nearest road for a couple of days would have really put us behind schedule! I also spent some time pondering how long it would have taken to walk east to the Haul Road! (Answer: Probably a couple of long, hard days!)

From our good original strip, Dave flies Don back over to Happy Valley. The Bill's and I get to work moving the moose meat to the better gravel bar. It's smoother and longer; exactly what we need for a heavily loaded Super Cub. It's only a 20 minute pack from kill site to camp. We get all the meat back and get camp packed up. Dave gets everything and everyone flown out to Happy Valley in short order. Don and Doyle have caught a ride south with George and Phil.

My Happy Valley neighbor and guide protege, Len Macker, has had an unusual mishap. He accidentally landed on a shed moose antler on the airstrip at one of his remote camps. He has punctured the $1,500 fancy tundra tire. That puncture would have been bad enough, but just after the antler stuck in the tire, the tire slung the moose antler up into the prop! A very unusual, and yes, expensive, mess. Len beat, bent and filed the prop out, patched the tire and flew empty over to the big strip at Kavik. He has someone from Fairbanks bringing him a replacement prop. His mom, Val, has procured a new tire and is sending it to Deadhorse tomorrow. Len will be back in the air soon.

The following day we get our basecamp buttoned up and make the long journey back to Fairbanks. This expedition takes less than 13 hours, including refueling at Coldfoot and the Yukon Crossing. And we set a new record for my outfit on this rough road... 900 miles roundtrip of gravel road driving without a single flat tire!

Chapter 22

Walking Wounded At North Fork

It was the end of September 1979. Red Beeman and I had just finished up a real terrific season at North Fork. We had hosted two Lower 48 clients on the multi-species hunts; both had done really well. Jim and Rich were both in good physical shape and real decent "walkers." Both were good marksmen, even though neither had a lot of experience hunting big game. Both were willing to do what we asked of them as we went in search of the game they sought. As a result of their willingness to be guided, both hunters were ultimately very successful. In fact, my client, Jim, had taken a B&C caribou, a moose, a grizzly and a sheep, all in a 15 day hunt. One bright morning Jim killed a grizzly and a moose in less than 20 minutes! The hunters were happy, there had been no expensive equipment breakdowns and nobody had gotten hurt.

We were finished with the paying customers. When the plane arrived to pick up the clients, on it was my "deer hunting mentor" Don D'Amato of Key West, Florida. When I was in high school back east, Don had taken me under his wing and imparted a ton of information about white-tailed deer. How to know them, how to hunt them and how to cook them! Don had also been my Best Man at my wedding to Jan in 1973. Now I would have a chance to help him achieve the dream of a big game hunt in Alaska. Don would be hunting for a trophy caribou, and there were some nice ones around.

Don hopped off the Southcentral Air chartered Cessna 185 and we had a terrific reunion. I had not seen him since my wedding six years before. Red Beeman had headed to town for a while. He would be back for a few days. Red's son, Eric was coming out to camp to spend the winter trapping there on the North Fork. The clients were gone, but there were still tons of things to do. But the first order of business was to find a big caribou bull for Don.

Walking Wounded At North Fork

Don and I spent the first night in main camp. Then we packed up some gear in the old Coot (articulated 4WD vehicle) and headed off toward the Middle Fork of the Kuskokwim where I had been seeing plenty of caribou and in fact, where my client Jim had taken his Boone and Crockett caribou not a week ago. It wasn't a long journey for the Coot. Perhaps a two hour jaunt. We got to the spike camp where Red and I had erected a wall-tent and put in a small wood stove. There were no bunks, but there was a delightful smelling spruce bough bed that felt heavenly when we turned in that night with the wood stove crackling away.

The following day was cloudy and overcast, but there was no snow or rain. We walked on over to the Middle Fork and peered into the drainage. It was flat "covered up" in caribou! We could literally see 100 caribou at any given time. The caribou in this area are not really migratory, so normally we do not see large aggregations of them. In this drainage, we normally see small bachelor groups. This was a very large group of animals, with an array of ages and sexes. And bulls? There were probably 30 bulls in the valley. They ranged from yearlings to old "white necks." We set up our spotting scope and started looking them over.

Making a choice was difficult! There were several bulls that were trophies of above average quality. But we kept coming back to a bull that was obviously getting up in years. He had a group of cows with him and he was obviously the herd master. Smaller bulls would come wandering by, but as soon as the old bull noticed them and started in their direction, they would skitter away across the tundra. No other bull in the valley wanted to test the dominant bull. This animal had a magnificent rack of antlers. He would be our target. Our most pressing problem became a function of avoiding so many alert caribou eyes in the valley. Not everybody had their heads down grazing all at once. Some caribou were always looking our way. There was just no way to get within shooting distance without being seen. We pondered this problem of "too many caribou" and came up with a plan. We would try just "moseying" into the valley, working closer and closer without walking directly at the bull. With luck, he would remain preoccupied with his harem and not become overly concerned with us until we got much closer.

Walking Wounded At North Fork

Don and I stood up, raised our arms above our heads so that we would hopefully look like caribou with antlers. We began our first diagonal mosey. We were relieved to find that it worked. Over 25 minutes, we worked within 350 yards of the herd bull without alarming him or any of his girlfriends. A few more minutes and one more diagonal mosey and we were inside of 200 yards. Now a couple of cows were nervously watching us. It was now or never. The mosey charade had worked admirably so far, but it was not going to keep on working forever.

A shot of 190 yards was still going to be quite a poke for a dyed in the wool brush country deer hunter from back East who was accustomed to most shots being way under 50 yards. But Don was a good shot and an experienced hunter. He had my 7 mm Rem Mag rifle and a good scope. I felt it was time to shoot. I set Don up with my pack for a rest. He "got on" the bull and watched for a while to be sure he was comfortable and that there were no other caribou behind the target animal. I was pretty sure he was ready, but was still a bit surprised when the shot rang out. The big bull took only a couple steps and toppled over. The fun part was over and the work was about to begin.

The rest of the herd drifted off down the Middle Fork. They really didn't spook, but rather just moved off. Some of the satellite bulls were probably pretty excited that we had just eliminated their main competitor. Perhaps they were getting ready to divvy up the cows?

We walked up to the fallen monarch. What an exceptional bull. In fact, it was and remains the largest caribou bull I have ever guided anyone to, much less taken myself. I was not carrying a tape measure, but I was anxious to get back to camp and slap a tape on this rascal. That would come with time.
We took a number of photos of Don smiling up from between those gigantic antlers. Then we spent an hour and a half butchering the bull and getting him ready for the long haul back to camp. I took most of the meat. Don took the rest and insisted on packing the antlers himself.

Back at the spike camp I green scored the bull. He was at least 425 inches of gorgeous antlers and would obviously go "way up" in the B&C record book. Don was ecstatic about his trophy. And well he should be. When the bull

Walking Wounded At North Fork

Don with his magnificent Boone & Crockett caribou bull on a delightful hunt that evolved into a painful nightmare for me.

was finally scored by an official B&C scorer, it went into the "book" at 432 inches and change!

We had fresh caribou liver with bacon and onions that night and we slept real well. Don was treated to the sound of wolves howling too. The next morning we slept in a bit, got a late start and headed back to main camp, stopping along the way to kill a few ptarmigan for the pot. We were back at main camp on the North Fork by mid-afternoon.

An hour or so after we pulled in, Southcentral Air was circling the airstrip with two Cessnas, a 206, and a 185. The 185 set down first. In it was Eric Beeman. The belly-pod was full of traps and the rest of the cargo area with seats removed, was filled with six airsick, puking, sled dogs. This was a heavy load; the pilot wanted to use all the strip. As the aircraft touched the gravel and the tail dropped down, the tail wheel caught in the last willows on the

Walking Wounded At North Fork

First order of business was to unload the dogs from the 185 and get them tied up. Further business is what became so exciting (and downright painful.)

downstream end of the strip. A bolt gave way and the tail wheel itself slipped out of the forks! Now, instead of having a rotating pneumatic tire and some steering, all the plane had was the forks dragging in the gravel. And of course it still had a belly-pod full of traps and, inside, the stout young trapper and his six puking sled dogs. The 185 skittered down the strip and peeled part way off into some short willows and stopped. With the 206 still circling the strip, we had to get the 185 out of the way.

Everyone realized it would be easier to push the 185 out of the way if we first disgorged the traps, the trapper and the puking sled dogs! Eric jumped out and staked out a "gang chain" off to the side of the airstrip in the willow brush. We then secured the six dogs to the gang chain and grabbed the traps out of the belly pod. Its load lightened considerably, we could now lift the rear fork and maneuver the aircraft far enough off the strip that the 206 could safely land. And land it did. The dogs were secure, so we concentrated on off-loading Eric's winter supplies from the 206. Once that was accomplished, we formed a skirmish line and began combing the willow brush at the downwind

Walking Wounded At North Fork

end of the strip where the tail wheel had last been seen. Twenty minutes later we found that missing tail wheel. Red had enough of a hardware collection that we soon turned up a bolt that would hold the tail wheel on the 185 well enough to get back and land at Merrill Field in Anchorage. The temporary repairs were made and both the pilots took off. We went back to the cook tent to reminisce. It had been quite a day.

As it turns out, it was a good thing that Southcentral Air had made it in. The next day we started getting rain and snow and fog. It stayed that way for the next few days. We stayed pretty close to the stove and performed "indoor" chores, including preparing the meat, cape, hide and antlers of Don's caribou. By the third day of crappy weather, we got to feeling pretty sorry for the six dogs tethered to the gang chain out by the strip. It was agreed that we would move them in under some big white spruce where they would be drier and hopefully happier.

We headed out to the open area near the airstrip where the bedraggled sled dogs were tethered in the brush. Since we needed the gang chain itself, the decision was made to move all six dogs at the same time, while they remained hooked up together on that chain. As it turned out, this was not a terribly good decision. In fact, in retrospect, it was a terrible choice. Eric got in the front, Don latched onto the approximate middle and I brought up the rear. To ensure a solid grip, I cleverly wrapped the end of the gang chain around my left wrist. We unfastened both ends and started out to the airstrip.

About the time we got to the strip itself, all six sled dogs simultaneously decided they were close to freedom, or perhaps going for a run in the snow. At any rate, their air sickness completely gone, all the pent up energy they had built up over the past three days kicked in. They shifted into high gear. At the front of the "pack" Eric let go when the first dog caught up to him and passed him. Don was jerked off his feet and fell to the side. That left me, my wrist firmly held by the end of the gang chain wrapped around it, as the only possible impediment to this miniature Iditarod gang start. I ran about three or four steps, but then I tripped and fell. When I hit the ground, I pitched forward and hit on my left shoulder. Something popped. Six dogs were dragging me down the airstrip, but all I was aware of was the excruciating pain

Walking Wounded At North Fork

Don D'Amato showing his shotgun prowess on ptarmigan during our Coot ride over to caribou camp.

in my shoulder.

After being dragged a few yards, the chain mercifully came unwrapped from my wrist. The dogs disappeared down the airstrip. I lay in the middle of the strip, looking up into the rain and snow, but all I saw was stars and streaks of bright pain. I reached around with my right arm and clamped onto the general area of where my shoulder should be. Where in normal times and up until a moment ago I would have felt a bulge where the humerus should fit snuggly into the socket on the scapula, instead of a bulge, I sported a hollow spot. I had suffered a very traumatic dislocation of that joint. My humerus (upper arm) had been pulled out of the socket. And when the pressure was released, instead of popping back into the socket, the head of the humerus

Walking Wounded At North Fork

had extended back up under the socket of the scapula.

Eric and Don came running up. The dogs didn't give a crap about me and continued off down the island. I showed Eric and Don the mess that had been my shoulder for the past 30 years. It was a mess. We were 90 miles of solid, roadless, wilderness from even the most rudimentary rural health clinic and a health aide. We were 225 miles from the nearest hospital or ER. I was in quite a pickle.

As it turns out, the level of medical assistance available is relative to where one is lying, injured on the ground, in a blizzard. In this case, Don, who had captained a fishing boat in Florida for many years, had seen a deckhand suffer a similar traumatic dislocation of a shoulder one time at the fuel dock. This guy was leaning out over a boat and lowering a boat motor. No dog team was involved, but there were similarities in the injury itself. Don had watched a fishing client/doctor "reduce" the dislocation and ease the humerus back into its socket. And, boys and girls, Don, having "seen a doctor do it," was at this time on a snowy September day in the Alaska Range, far from professional care, the finest medical help available. Don reached down, grabbed my wrist, planted his foot on my chest and began trying to rotate the bone back into the socket. As much as I would like to describe to you how this worked, the fact is that I have absolutely no idea. I screamed like a little girl and promptly passed out. I awoke to Eric slapping my face and asking (get this....) "Pete, are you OK?" I have never, before or since, experienced such pain and agony. Clearly I was not "OK!" "Damn Eric....! Do I in any way appear 'OK?'"

The guys helped me to my feet and supported me as I staggered 100 yards to the cook tent. This was some five-star, Type 1 pain.

Eric at some point came up with the brilliant thought to get out the "Wilderness First Aid and Medicine" book that we kept by the radio. He flipped through it and read aloud. The first thing the book mentioned was that this injury would cause tremendous pain. Check! That fits the scenario perfectly. Next it cautioned "Never try and reduce the dislocation yourself. Get the patient transported to an Emergency Room as soon as possible." Check... OK, never mind...! The book went on to detail all the terrible ramifications that could result if rank amateurs tried to get the bone back

Walking Wounded At North Fork

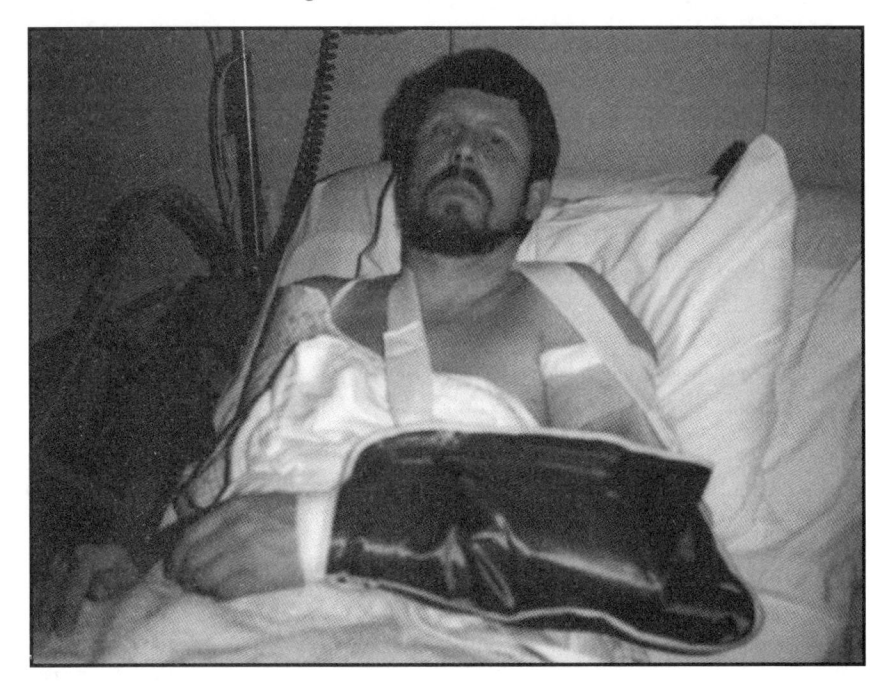

Eventually I got back to McGrath, then Anchorage and finally into the ER at Fairbanks Memorial Hospital.

into the socket. "It is very easy to pinch off blood vessels or nerves..." The one useful tip was that we should immobilize the arm. Eric and Don teamed up to put my arm in a sling and then wrapped ace bandages around the arm in the sling, binding it to my upper torso.

At this point, my number one priority was to try and find some way to deal with the excruciating pain. The book was not helpful in this regard, so we inventoried our "supplies." What we found was a couple of containers of aspirin and a bottle of Scotch. That relatively benign combination was the finest pain control available at this point in time at this location. I was reduced to doing the best I could, on a diet of aspirin and Scotch, for the next two days while the weather raged outside the tents. I was in a LOT of pain.

Walking Wounded At North Fork

By the time the storm began to let up, we made HF radio contact with an air taxi in McGrath, 90 miles away. They promised to come pick me up as soon as the weather would allow. It would have to be good both at McGrath and at our camp. Hopefully the area in between would be flyable as well. And so when things were looking better on both ends, the air service sent a plane to pick up Don and me and fly us into McGrath. In those days, Wien Airlines flew a jet in and out of McGrath just one day a week. Coincidentally, the day the weather cleared and I was taken into McGrath was that day! We landed in McGrath just an hour before the once-weekly jet was due to fly out back to Anchorage. We bought tickets and hopped on the plane. Once in Anchorage, we considered going to a local emergency room, but instead decided to catch a flight home to Fairbanks. Before leaving Anchorage I called a physician friend in Fairbanks. Rich was at a cocktail party but he agreed to meet us at the ER at Fairbanks Memorial Hospital (FMH.)

When we landed in Fairbanks, we went straight to FMH and shared our story and my physical mess with Rich. He ordered X-rays and examined the site of the injury. More to the point, he gave me a massive cocktail of pain medicine. In the course of a couple hours at the hospital, Rich determined that despite my amateur medical attention, I did not seem to have pinched off any blood vessels or nerves. He warned me that I was going to be in for a ton of physical therapy before I regained use of that arm. Boy was he right! I went home that night, and thanks to the pain meds prescribed by Rich, I did not wake up for a day and half!

My shoulder slowly healed and I began physical therapy. This involved a whole new level of pain! I went through 18 months of physical therapy before I was able to lie on my back and raise my left arm to a completely upright position, perpendicular to my torso.

Don spent a few days in Fairbanks playing with his honorary grandson and then returned to his home in Florida. For many years his caribou skull with its Boone & Crockett class antlers sat at the edge of his pool. It was utilized as a rack for hanging barbeque utensils!

Chapter 23

Buckaroo Stage

In 1990, I met an interesting cowboy character from New Mexico. Patrick Lyons had a family ranch (cow-calf operation) at Cuervo in east central New Mexico. He worked the outfit with his brother. The hard work of ranching was apparently too mundane. Pat had some pet issues and at some point decided to make a difference by running for the New Mexico Legislature. Surprising no one, he was elected as a State Senator.

Pat had swapped some hunts with Dave Bridges and me and he had been to Alaska a time or two, taking sheep, grizzly and other trophies. As with many of our swap hunt buddies, it got to the point where none of us could recall "who owed who" how many hunts. But we kept on hunting with each other. We just no longer kept score.

New Mexico has a system of "landowner permits" for most of its big game species. Naturally, when I decided it would be fun to hunt the trophy pronghorn antelope that New Mexico is famous for, I called Pat. I suggested perhaps some time when he had a couple antelope permits for his ranch, my son Jason and I would like to buy them. It didn't quite work out exactly the way I envisioned. It was less than a week later that Pat called and said he had landowner antelope permits for us. And by the way, they were not for his ranch, but rather for a huge spread in northeast New Mexico, just south of the Colorado line near Raton.

Further complicating the situation, the hunt was scheduled to start in just a week! I hastily booked airline tickets for Jason and I to fly to Albuquerque. Pat met us at the Sunport and we drove north to meet the Van Sweden family of the V-7 Ranch at Raton. It was the beginning of a terrific friendship and fantastic experience learning about ranch life. The buckaroo/cowhand stage of my upbringing was off to a great start.

Buckaroo Stage

Unlike most antelope hunters, we did not swoop in, shoot a couple of antelope and then fly home. I spent evenings in camp discussing with the Van Sweden family what they planned to do in developing their own outfitting business and working it into their cow-calf cattle operation. It seemed complex, but doable. The Van Sweden's knew a ton about cattle ranching. I had a fair grasp of outfitting and guiding. I also had a mailing list of a lot of hunters I had guided over the years in Alaska. I surmised that several of them might be interested in hunting Rocky Mountain elk in northern New Mexico. Turns out I was correct in this regard.

Jason and I had a terrific antelope hunt with the Van Swedens. Jason shot a "Booner" with a modern rifle and I took a very nice buck that easily made the muzzleloader book, "The Longhunter." In the case of my muzzleloader buck, no one was more surprised than I when, as the smoke cleared, there was an antelope lying dead on the ground! Jas and I returned to Alaska with our antelope and some great new friends. I made arrangements to trade Jason Van Sweden a moose hunt for a late season elk hunt on the V-7. That swap worked out well. I got a V-7 elk and Jason took a nice Alaska bull moose.

The V-7 had been maintaining a deal with a neighboring property whereby they utilized the adjacent ranch for "summer country" for grazing their cattle. In return, the neighbors received most of the V-7 landowner elk permits and hunted there. At some point this deal soured when the neighboring landowner wanted less cows on their ranch in summer, but still wanted the same number of elk permits from the V-7. The arrangement fell apart. The V-7 made the decision to utilize their own landowner permits and to guide elk hunters. The plan was to use the proceeds from the guiding to make payments on another chunk of grazing land, suitable for summer grazing. The new place was at Maxwell, New Mexico, a few miles south of V-7 headquarters. Van Sweden family members would do most of the guiding, but they could use another guide and I was selected. Suddenly I was not only a sheep-caribou-grizzly-moose guy, I was also going to be an elk guide.

Books have been written about what I didn't know about elk. Sure, I had killed some elk. I had taken elk in several states including of course, right there on the V-7. But could I hack it as an elk guide? The Van Sweden's seemed to think so and they knew a lot about elk! After some introspection,

Buckaroo Stage

When they are working cattle at the V-7, its all hands on deck, including ladies and children.

I decided that it was worth a try. In my opinion, the main thing I had going for me was my firm belief that the most important part of guiding was the manner in which you communicate with your hunters and take care of them. Beyond that, I knew a fair amount about elk habits and habitat. The elk were there on the V-7; it would all work out.

Once the animal is down, skinning an elk isn't much different than skinning any other big game. Soon I was to learn that skinning and butchering elk at the V-7 was even easier than the meat and trophy handling I was accustomed to. Most places that elk were killed on the V-7 were accessible by pickup truck, or occasionally via horse and then pickup truck. A small flatbed trailer was often utilized to transport the fallen elk back to a sturdy meat pole at ranch headquarters. There a bucket on a farm tractor raised the animal into the air. The innards were tumbled out into the bucket and finally, the entire skinned carcass placed on a clean tarp in the back of a pickup truck and driven to the processor in Raton. Easy peasy!

Buckaroo Stage

Add to this there was no "camping out" involved! I slept in the bunk house and ate fabulous meals cooked by the divas of the V-7 kitchen, Misty and Vivienne! I learned to love Tex-Mex cuisine. From the ladies I picked up awesome recipes and secrets of southwest cooking that serve me well to this day in my culinary endeavors.

Two other things I picked up while becoming an elk guide in New Mexico were a couple of extremely important safety-related truisms. First, when dining in New Mexico do not believe any cowboy when he says "Go ahead and eat some of that... It's not hot!" The practical translation of this innocent sounding phrase turns out to be: "Watch the gringo bite into that and get a mouthful of chilies! He will light right up!" Many was the time, before I smartened up, that I stuffed innocuous appearing southwestern cuisine into my pie hole only to a moment later be making a dive for a glass of water or milk just to survive! Of course the "New Mexico State Question" remains: "Red or Green?" In any restaurant or dining room, you are expected to be clear as to whether you prefer your meal smothered in green chili sauce, or red chili sauce. "No chili sauce" is simply not an option in terms of tradition, culture and good manners in the Land of Enchantment!

While we are on the subject of "things not to believe a New Mexico cowboy when he says...," I hasten to add there is yet another one that has the potential to kill you. Do not believe any New Mexico cowboy who hands you a lead rope and a saddle and says "Take that horse there in the corner.... He's gentle!" The translation of this phrase from the lexicon is this: "Watch this... that horse is going to buck that northern guy off so fast that his ass won't catch up with him for a couple of days!" It should have been a clue for me the first time I saddled up, when all the real cowboys stopped what they were doing and started giggling and pointing at me!

Hunting and my friendship with this ranch family had drawn me to northern New Mexico. Some of that hunting involved the use of horses for transportation, as well as for guiding and cattle work. Back in my misspent youth in Sussex County, New Jersey, I had literally ridden horseback before I was able to walk. My mom's side of the family had always been horse people. Much of my mom's life as a veterinarian revolved around horses. As we grew

Sampson the Devil Horse, tied to a juniper tree during a calm period in gathering cattle.

up, my sisters and I had horses. We had all taken riding lessons. We all had 4-H horse projects. But those show and even trail horses back East were not what I was dealing with in the mountains of New Mexico. These were work horses, not pampered pets. Some had been raised there on the ranch. Others came from the auction block. On the V-7, horses were one more "tool" in the cattle ranching tool box. The only exception were the ponies the ranch kids rode. They were docile by nature. They never bucked; not even the occasional crow-hop. Unfortunately, my legs were too long for me to ride these tame and thus safe creatures.

So I was destined to be riding "work" stock that were, as it turned out, in many cases much tougher than I was. It also seemed I was destined to keep proving that I was not as tough as my new equine friends. It also seemed that I was destined to provide a never-ending source of entertainment for the

Buckaroo Stage

I was literally riding horses before I could walk. I learned later in life that not all horses were as gentle as my Mom's pets!

real cowboys. At some point, I became so enamored with cattle ranching and the V-7 that I began volunteering to stay a while after hunting season to "help" with gathering, moving, sorting, doctoring and handling cattle. Oh, for the life of a cowboy. This would be fun.

I have always been a fan of rodeos. I love to sit in the stands with the other "non-athletes," cheering when someone makes a good ride on a bronc or a bull and groaning when a cowboy comes unstuck before the eight second buzzer goes off. At one point I queried some of the V-7 cowboys if they ever competed in rodeos. "Hell no" was the answer. "Those guys only have to stay on for eight seconds!" Granted, they had a good point. To someone with a few more IQ points than I obviously possessed, it would also have been a clue of what was to come as I charged headlong into the world of working cattle on horseback. I, on the other hand, completely missed the obvious clue.

Buckaroo Stage

The V-7 cowboys gathering bulls to be moved from the Bull Pasture down to the river bottom.

For some particularly unlucky reason, my first foray into gathering cattle came as we attempted to move the resident "grumpy old men" (Simmental and Hereford bulls) from the Bull Pasture to an area of lower elevation, much closer to ranch headquarters. By the end of the first morning of bull gathering, my visions of a bucolic cattle drive were pretty much in tatters. I spent a long day in the saddle, did NOT get bucked off, and managed to survive in one piece. Over supper I realized that I was pretty dang tired, especially considering that my horse had done all the hard work.

Perhaps my horse was a bit resentful too. The next morning we saddled up, loaded our trusty horses into a trailer and headed out to gather more cattle. After unloading and checking the cinches, we began mounting up. My equine nemesis, a lanky bay gelding named "Sampson" clearly had trouble on his mind from the onset. As I swung aboard, Sampson began to "crow hop."

Buckaroo Stage

Shouting "whoa" had no effect on his antics. Nor was it especially helpful that I had only my left foot in a stirrup. Sampson tested me; I failed the test. Not only did I come loose fairly quickly, but he managed to strategically launch me so I came down astraddle a barbed-wire fence. It was pitiful and pathetic. On the entertainment scale, it rated at least a 9.5 out of 10. I imagined the crew rushing to my aid. Instead I found I was once more the object of entertainment and pretty much on my own! I rechecked the cinch and resolutely remounted. This time I was ready for another minor rodeo. When Sampson began to bounce, but I quickly jammed my right boot into the off-side stirrup, reefed up the reins and pulled his head up. I thought I detected a modicum of respect from Sampson. In retrospect, the diabolical horse was probably just slacking off to get a better hold!

We brought in quite a few bulls and got them corralled in the working pens to be wormed and doctored. At noon we even drove up to town to munch on cheeseburgers (with the obligatory green chili) at the local Dairy Queen. I swaggered into the Raton Dairy Queen in my cow-shit splattered boots and chaps, with my spurs a-jingling. The tourists who had stopped in from Interstate Route 25 were definitely impressed at lunching with "real" cowboys. As long as the cowboys stayed on the downwind side of the Dairy Queen dining area. The tourists had no way of knowing that the V-7 ranch girls were real "cowboys;" I might have looked and smelled the part, but I was merely an imposter in cowboy getup!

After lunch we headed back out and I found myself working some bulls without any help. Sampson and I found a pair of Simmental bulls off in a gully. We hazed them out and pushed them over toward the pens without much drama. But once these boys got to a little open country, they decided they liked it better back in the brush where they had spent most of the fall. The pair swapped directions and headed back to their lair. There was only a minor obstacle blocking their way... me and Sampson! As I quickly learned, for a couple of 1,500 pound bulls, a bay cow horse and an Alaskan "imitation cowboy" do not constitute much of an impediment to escape. Instead of being impressed (much less scared of) of my cowboy type noises, the bulls came right straight through us. Sampson probably forgot I was on his back. His primary objective now was simply to get out of the way of the oncoming

Buckaroo Stage

New Mexico version of two Cape buffalo. He rapidly lurched sideways which put me off-balance in the saddle. Once those "load arrangement" and "weight and balance" issues cropped up, Sampson's next objective was to try and rid himself of me so that he could get back to the business of avoiding being run over by the bovine monsters. It did not take him long. He began to seriously buck and I came loose. I would have been smarter and better off just to bail out of the saddle and be done with it. But I had left my "smarter" and my "better" thinking back at the truck. Instead I tried to regain my seat. That was not likely going to happen. I came out of the saddle a time or two and each time bounced straight back down. Not "down" as in settled gently into the saddle either. Rather, my crotch crashed into the saddle. It was all I could do to avoid straddling the saddle horn itself. Even that avoidance strategy did little good. I soon came completely loose from my horse. Sampson ran a few yards and stopped (bless his heart...) The bulls' butts disappeared into the thick brush of their home coulee.

Truth be told, a lot of me was just numb. Especially some rather important appendages in my nether regions. I tried to stand up and had a ton of trouble doing so. I tried crawling over to my horse, but Sampson had suddenly developed a distinct level of distrust of a "cowboy" on his hands and knees muttering miscellaneous profane invectives. At first, any time I got close to him, he went all wall-eyed and shied away. Just out of reach. I suppose I should have been grateful he didn't shift into high gear and barrel off all the way to the barn. He and I were still on the same half-acre of pasture; I still had a chance.

Finally I crawled close enough to grab one of the reins that he was trailing on the ground. With that leather firmly grasped I spoke soothingly to Sampson. Somehow I seemed to convince him we would just leave those two bulls right there in their hidey hole and go back to the truck, if only he would hold still long enough for me to climb aboard. It took a few minutes but with the aid of a nearby juniper tree I was able to clamber aboard my skittish horse-friend. The "fires of hell" sensation in my crotch was a constant reminder that I was not a cowboy and should NOT be riding a horse. Yet, the alternative, walking back to the trucks, held even less appeal. I estimated the trip back to the trucks and trailer felt like 100 miles and two days in length. In truth was probably a twenty minute ride. Not surprisingly, there

Buckaroo Stage

was STILL no sympathy from the assembled cowboys. In fact, I was now the object of even more derision and the butt of even more jokes. I didn't care; I was back at the truck and all I wanted to do was to go lie down. If the last horse suddenly disappeared from the face of the earth, that would have been a good thing. I also had visions of Sampson being recruited for future service either as glue or dog food.

I crawled into a truck and did my best to elevate my crotch as we bumped along the two track to the barns. Once there, I completely shirked my cowboy duties and responsibilities. I gimped into the house to retrieve a bag of ice. Some other kind soul unsaddled Sampson and turned him out. An hour later it was clear that I probably should have a trip to what passes for an Emergency Room in Colfax County, New Mexico. I talked Jason Van Sweden into hauling what was left of me up to the Colfax Clinic at Raton. Not surprisingly, the staff there were all "horse people." In fact, they showed a bit more sympathy for my plight than had the ranch hands. After some painful prodding and poking and a hypothesis that I might have a separated pelvis, I was wheeled off to the x-ray machine. The films showed that I did NOT have any broken bones or even a separated pelvis. The ultimate diagnosis was just severe muscle strain in my general "crotch" area. I was prescribed some heavy-duty pain meds and told to "rest up" for a few days. Yippee..... I was out of the cattle wrangling business now!

The next problem was that I was due to drive to Albuquerque and fly home in two days. On doctor's advice, I changed my reservation and remained horizontal at the ranch to heal for a few more days before returning to Fairbanks. At no point during my remaining time on the V-7 did I attempt communication or contact with Sampson. Somehow I doubt he was concerned about me either!

In a month or so I was completely healed up and able to walk a straight line without gimping. As much fun as it had been, apparently my cowboy days were now behind me. I had been riding horses, and for that matter, getting bucked off of horses, all my life. What I learned at the V-7 that fall was that as I had aged, getting back up off the ground was not as easy as it had been years ago.

End